AF442057

STATISTICS FOR EVERY FAN

Sports Statistics
SPRINGFIELD COLLEGE

Carl Fetteroll

Springfield College, Massachusetts

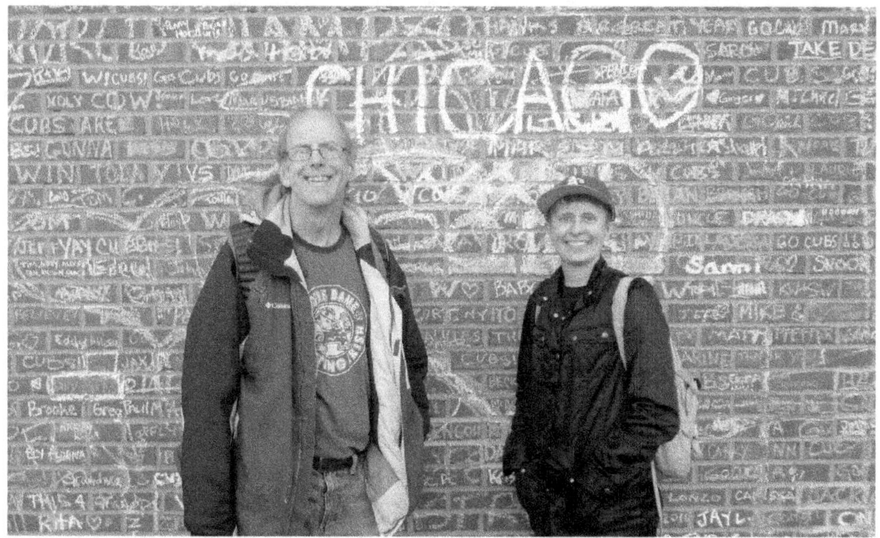

Wrigley Field, Chicago, November 3, 2016, the day after the Cubs clinched the World Series Trophy.

Carl Fetteroll

Author

Carl Fetteroll has been an adjunct math professor at Springfield College in Massachusetts for over 20 years. Carl is a past coordinator of the Massachusetts Senior Games and director of the recreation program for children with visual impairments at Springfield College.

In his spare time he participates in triathlons, completing two Iron Man distance triathlons, the Great Floridian in 2014 and Ironman Quebec in 2015.

Danny Fetteroll

Illustrator

Danny Fetteroll is an illustrator and blogger for the Oakland Athletics branch of SB Nation. Born in Boston he has had an odd obsession with the Oakland Athletics since he was a child. To pay the bills he works as a barista which allows him the flexibility to attend plenty of baseball games.

Danny's Athletics comics and player profiles can be found at sbnation.com/users/fatrolf/blog.

History of Sport Statistics class at Springfield College

In 2005, the math department at Springfield College discussed how to make our elementary statistics course more accessible for our students. Most were taking Algebra assuming it would be easier than stats, since they had all taken it in high school. Plus statistics was thought of as indecipherable and with the huge $200 textbooks filled with problems that students had no interest in, impenetrable.

Looking at our undergraduates, 92% had played sports before coming to Springfield College, so I suggested we change our elementary stats course to a sports stats course, teaching the same material in a sports context. At the time, there were no textbooks with 100% sports examples and problems. So I decided to write the book for the class.

The book that follows is the one I wish I had in college. Because of the focus on sports, I had an added benefit. I was able to cover more statistics than a college-level stats class. I added chi-square and ANOVA, and we are able to do confidence intervals and hypothesis tests for 1- and 2-proportions, and 1- and 2-sample means. The other benefit was the ability to teach the class in a computer lab, where we use Minitab® Statistical Software for every analysis we do.

Since I was able to design the course around the book, Springfield College now has one of the first college-level stats classes following the guidelines developed by the American Statistical Society in 2003. The Guidelines for Assessment and Instruction in Statistical Education (GAISE) focus on introductory college courses, like ours. The goal is for students to be statistically literate upon completion of their course. The GAISE recommendations are included in this book and our course. They are:

- Emphasize statistical literacy and develop statistical thinking;
- Use real data;
- Stress conceptual understanding rather than mere knowledge of procedures;
- Foster active learning in the classroom;
- Use technology for developing conceptual understanding and analyzing data;
- Use assessments to improve and evaluate student learning.

At the conclusion of the course, my goal is to use the above guidelines to insure this class produces statistically educated students through development of statistical literacy and ability to think and interpret statistically. I hope you enjoy your journey through the amazing world of sports statistics!

Note on Minitab Statistical Software

For years this course was taught using the TI 83/84 family of calculators. Then, in 2014, we decided to use the computer labs at Springfield College for our sports stats classes. That meant I could stop requiring a relatively expensive calculator that was a one-shot purchase for the elementary stats students I had in class. For most, this would be the last math class of their college careers. Instead, the calculations could now happen using available software.

At Springfield College, our choices were Minitab, Excel, and SPSS. Each can easily do the statistical calculations required in Sports Stats, but Minitab was by far the best choice for our students. Minitab was created at Penn State in 1972, and the designers are also users. They knew what makes it easy for students (or conversely, frustrating for students). They designed a very easy to use, and for me, easy to teach program that is very intuitive for students.

Minitab is the only software program that allows me to teach the entire syllabus of elementary statistics (plus chi-square and ANOVA) while the students are simultaneously learning Minitab. With zero frustration on the students' part. We can also take advantage of the incredible wealth of sports data on the internet in the blink of an eye. Every problem I do as I teach, and every problem the students do on their own while learning, is real data with real results, and no frustration. That creates an atmosphere where learning can take place, which is all I ask for! Thank you, Minitab!

Carl Fetteroll

Springfield College, Math, Physics, Computer Science Department

Springfield MA

August 25, 2020

Introduction to Sports Statistics

We will cover sports statistics in three parts.

Part 1 is Descriptive Statistics

Descriptive statistics describe or summarize what happened in the past. Tuukka Rask's save percentage for the current season. Or for his career. The Celtics 3-point percentage up through today. Or for the current homestand. Or their current head coach. Tom Brady's career stats with the Patriots. Not his future stats with the Bucs. Their *past* data.

With descriptive statistics we can turn a whole bunch of numbers (data) into a single number to **describe** something.

It's what a good journalist does. They talk to lots of people and COLLECT a big confusing pile of data. Then they ORGANIZE the data, ANALYZE it to give us an INTERPRETATION (snapshot) of what has happened. There's no prediction of what will happen next. There's no guessing about why. Just a simple picture of what is, right now.

We can use Descriptive Statistics to answer questions like "How many home runs did Mookie Betts hit in the 2019 season?" We can gather data from each game and add up all the home runs to get a single number. We can turn every MLB player into a single number, then compare them to each other to answer, "Who were the top ten home run hitters in 2021?"

Part 2 is Predictive Statistics

Predictive statistics helps us guess what might happen next. Tuukka Rask's save percentage next year based on this season. Tom Brady's completion percentage next year with the Bucs based on his last season with the Pats. The Celtic's 3-point percentage next year based on this year's stats.

With predictive statistics we can use past data to **predict** future results.

It's what a weather forecaster does. They use data from what's happening now and what happened in the past to predict what might happen tomorrow.

We can use predictive statistics to answer questions like, "How many home runs will Alex Verdugo have with the Red Sox in 2020 based on the home runs he has hit throughout his career?" and "How many touchdowns will Tom Brady throw in 2020 with the Bucs based on his career with the Pats?" How many games will the Red Sox and Celtics win *next year* based on how many games they won *this year*?

Part 3 is Statistical Inference

In Part 2, we use past data to predict future performance, but our prediction lacks certainty. In Part 3, we will test the samples two different ways to look for certainty.

We will be detectives. We will use a sample to deduce — to **infer** — what is going on. Then test it two ways. Inference uses reasoning to draw a conclusion based on evidence (our sample).

We can use statistical inference to answer questions like, "Were NFL kickers worse on extra points in 2015 after the NFL moved the extra point back 13 yards?"

By the end of the course you won't be able to look at sports statistics the same way again.

Contents . . .

Statistical techniques are tools of thought, not substitutes for thought.

— Abraham Kaplan, American philosopher

PART 1
Descriptive Statistics

A passion for statistics is the earmark of a literate people.

— *Paul Fisher, Fireside Book of Baseball*

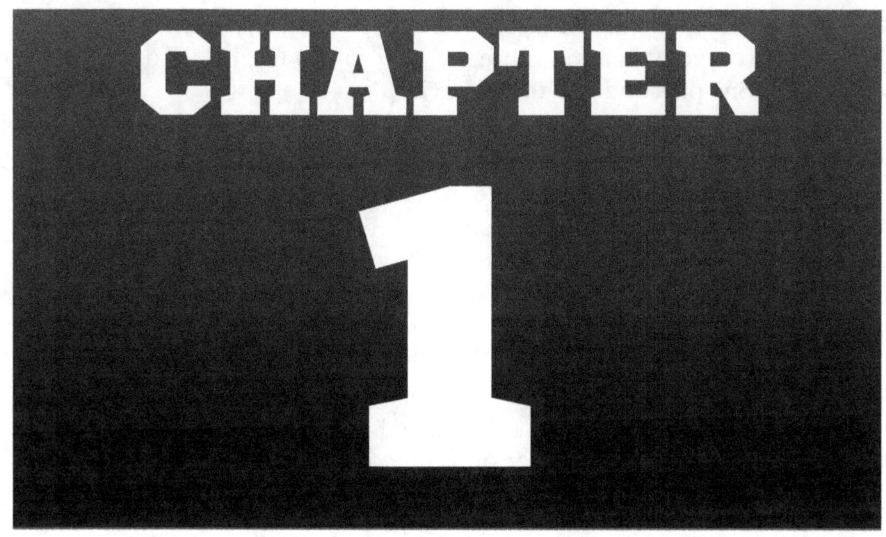

Brady. Brees. Manning.
Who is the best quarterback?

It took me seventeen years to get three thousand hits in baseball. I did it in one afternoon on the golf course.

— *Hank Aaron, Home Run King*

PREVIEW: TURNING TOUCHDOWNS, COMPLETIONS, YARDS INTO ONE SPORTS STATISTIC

The NFL season was at its half-way point in 2009 when its three marquee quarterbacks respectively led their teams to victories. Tom Brady, Drew Brees and Peyton Manning are familiar names even to those who do not follow football, and in week 9 of the 2009 NFL season they did not disappoint. Tom Brady and the New England Patriots beat the Miami Dolphins 27-17. Brady was 25 for 37 and 332 yards, with one interception and one touchdown. Drew Brees helped beat the Carolina Panthers 30-20 by going 24 for 35 and 330 yards, and he also had one pick and one touchdown.

Who had the better day? They both won and they both threw for over 300 yards. The rest of their stats were almost identical. That same night, Peyton Manning led the Indianapolis Colts to a come-from-behind-miracle victory over the Houston Texans to keep his team undefeated. The Colts won 20-17, and Manning was 34 for 50 for 318 yards with, you guessed it, one pick and one touchdown!

Those were three of the top performances of the week, but who had the best game? It's an easy question to answer when athletes go head-to-head in the same competition. For instance, at the 2010 Massachusetts State Track and Field Championships, three girls broke the state record in the 400M hurdles, and all had phenomenal performances. But they did it in the same race, it was easy to say who had the best day. With Brady, Brees and Manning, they were in different cities and playing different teams. How do we rank their performances?

The NFL has come up with a statistical method for measuring the performance of quarterbacks. It's called the passer rating, popularly known as the quarterback rating. (In college it's also called passing efficiency.) It's made up of 4 components:

$$C = \left(\frac{\text{Completions}}{\text{Attempts}} - .3 \right) \times 5$$

$$Y = \left(\frac{\text{Yards}}{\text{Attempts}} - 3 \right) \times .25$$

$$T = \left(\frac{\text{Touchdowns}}{\text{Attempts}} \right) \times 20$$

$$I = 2.375 - \left(\frac{\text{Interceptions}}{\text{Attempts}} \times 25 \right)$$

where

Attempts = Number of passing attempts

Completions = Number of completions

Yards = Passing yards

Touchdowns = Number of touchdown passes

Interceptions = Number of interceptions

If any result is greater than 2.375, it is set to 2.375. If any result is a negative number, it is set to zero.

The calculations are then plugged into the formula:

$$\text{QB rating} = \left(\frac{C + Y + T + I}{6} \right) \times 100$$

Tom Brady that day would be calculated as follows:

Attempts = 37
Completions = 25
Yards = 332
Touchdowns = 1
Interceptions = 1

$$C = \left(\frac{25}{37} - .3 \right) \times 5 = 1.88$$

$$Y = \left(\frac{332}{37} - 3 \right) \times .25 = 1.49$$

$$T = \left(\frac{1}{37} \right) \times 20 = .54$$

$$I = 2.375 - \left(\frac{1}{37} \times 25 \right) = 1.70$$

$$\textbf{QB rating} = \frac{(1.88 + 1.49 + .54 + 1.70)}{6} \times 100 = 93.52$$

Brady's quarterback rating for the game was 93.52. (For comparison, quarterback ratings go from 0 to 158.3 in the NFL.)

Try it! (primecomputing.com has a Quarterback Rating calculator if you want to play around comparing quarterbacks.) Then calculate the ratings for Manning and Brees and you should get 83.6 for Manning and 96.1 for Brees. Who had the better day? Drew Brees with his 96.1 surpassed Manning and just edged out Brady. Manning's rating was hurt because his completion percentage was lower, and Brees edged out Brady because he accomplished basically the same results but with two less attempts.

Was Brees the best quarterback that Sunday? Actually, no. Jay Cutler, in a loss to the Cardinals, was 29/47 for 369 yards and had 3 touchdowns and 1 interception. His three touchdowns pushed him to a 98.6 rating, the highest for a quarterback in the NFL in week 9.

Be careful how you use statistics though; don't rely on just the statistics. Use statistics to guide you in interpreting the data. Cutler had the highest rating, but his Bears suffered a devastating loss to the Cardinals, a game they should have won. Manning was trying to keep his team undefeated through 9 weeks and had 50 attempts because of the miracle comeback needed to beat Houston. The Patriots were trying to show the Dolphins that they were still the beast of the East. And Brees had his Saints undefeated, poised for a showdown the following week with Indy. Statistics provide us with a tool to help us interpret the game.

The quarterback rating is fine for distinguishing between performances in terms of numbers, but always ask yourself is there something else? Why did Manning throw 50 times? Sports statistics is great for giving us numbers, but don't stop there, we need to analyze and interpret those numbers; don't just accept them at face value. Use them as your starting point in your journey into the world of sports statistics.

PRACTICE PROBLEMS

1 Look at this weekend's games in the NFL. Pick two quarterbacks you think will have high quarterback ratings. (Remember, in the NFL quarterback ratings go from 0 to 158.3) Or your two favorite quarterbacks. Or the quarterbacks for your two favorite teams. Pick them now, before the games are played. That will make it more fun. After the games, compute the ratings for each quarterback. Their stats will be at NFL.com. NFL.com will also have their ratings calculated and you can double check your numbers. On Monday turn in your ratings for each quarterback — show all your work — and your reasoning as to why one was ranked higher than the other.

2 Do the same assignment as in Question 1, but for two college quarterbacks. For college you'll need a different formula. The NCAA calculates their quarterback rating differently from the NFL. Both use the same stats, completions, touchdowns, interceptions and average yards, but the NCAA system (which most high schools also use) uses a different formula to calculate the rating:

$$\frac{\Big((8.4 \times \text{Yards}) + (330 \times \text{Touchdowns}) + (100 \times \text{Completions}) - (200 \times \text{Interceptions}) \Big)}{\text{Attempts}}$$

The scale for NCAA goes from −731.6 to 1261.6. To put that in a better perspective, the 100 career leaders ratings are from 175.62 to 145.30. The weights (8.4, 330, etc.) were chosen in 1979 so that an average passer would have a rating of 100. Because of rule changes and improved passing, 100 is now a poor passer.

Comment on the differences in ratings between your two college quarterbacks. Why was one ranked higher than the other?

 In 2019, Springfield College had the nation's best rushing offense in all of Division III, rushing for 357.4 yards per game. Clearly the Springfield College football team does not rely on passing. But as we just saw, a quarterback's rating depends on passing. How does our SC run oriented style of offense affect the QB's ratings?

Calculate the quarterback rating for Springfield College on a recent random game. To get the data, go to springfieldcollegepride.com. Go to Athletics, pick Football, click More+ and pick Statistics, then select Game log. The score is an active link, allowing you to scroll down and click on the score of the game you are interested in. Game log defaults to the box score, just scroll down for the QB stats.

As an example, here's the box score for Western New England vs. Springfield College, 9/7/2019. And yes, you are reading the box score right, we were a combined 2-9 in that loss:

Springfield College

Passing	C-A	YDS	TD	INT
David Wells	1-6	20	0	0
Chad Shade	1-3	3	0	0

After you pick your game, then use the same formula in Question 2.

Does the quarterback rating reflect the type of offense they run? How useful is the quarterback rating in this case?

Turning numbers into data.
Is it magic?

I'm a numbers guy, and I think numbers sometimes tell stories and sometimes they don't. When you look at the NBA, when teams shoot 45% or better from the floor, what is their record? And if they shoot under that what is their record?

— Doug Collins, 76ers coach

PREVIEW: NUMBERS DO NOT EQUAL DATA. WE NEED CONTEXT

Defining data

30, 40, 93, 95, 142, 155, 175.

Numbers. Seven numbers specifically. Alone, they are just numbers. But do they have any meaning? We need context to understand their meaning. OK then, here is the context:

NCAA DIV 3 X-Country New England Regional Qualifier
USM @ Twin Brooks - Cumberland, ME
Saturday, November 14, 2009
11:00 AM (Women's 6000 Meter)

These seven numbers represent the seven women on the Springfield College cross country team who ran in the regionals. The numbers are their scoring place (i.e., Amanda DiPaolo of Springfield College finished in 30th place, and therefore scored 30 points for Springfield College). The top 5 runners for each team score, Springfield's team score was 400 (30 + 40 + 93 + 95 + 142), which put them in 13th place out of 47 teams competing.

Alone, 30, 40, 93, 95, 142, 155, and 175 have no meaning. But give them a context they make sense. They tell a story. The numbers have become data.

Data is the most important concept in statistics, never mind sports statistics. We will use data every day in class, just as we do in our daily lives. Turn on ESPN and be amazed at the data they dig up and present. Fantasy baseball and football are all about data. Read a box score one morning from a late night west coast game, and it is as if you watched the game. You can visualize the game taking place as you scan the box score. Look at the amount of data needed to calculate the quarterback ratings in Chapter 1. And the ratings themselves are more data for us to analyze and dissect. You can keep the entire lunch-table conversation going with one easy question, Who was the best quarterback of the first decade of the 21st century, Tom Brady or Peyton Manning?

Defining statistics

Now we can define sports statistics. This course is not about memorizing. I will never ask you to define sports statistics, or calculate a quarterback rating without giving you the formula. This course is about finding data and using data to analyze and interpret events in your life. We use sports examples in this class because we have all played sports and we can understand the examples.

You can apply these techniques to the rest of your life; the skills you learn here are transferrable to everyday problems. And the biggest skill is developing the ability (or the confidence) to ask questions. I will show you how to do the statistics, and you will get an answer. But then ask yourself why did that happen? Why was Drew Brees ranked higher than Peyton Manning? Analyze the data. Interpret it. In this class, we may have 25 different interpretations, don't worry if your interpretation disagrees with the teacher's. (It probably will!) The only answer that would be wrong is the blank one. So, process the data and don't worry. We will hear everyone's interpretations and learn from them.

Given that, let's define statistics. **Statistics** is a branch of mathematics that collects, organizes, analyzes and interprets data. Sports statistics specifically deals with **COLLECTING, ORGANIZING,**

INTERPRETING and **ANALYZING** sports data. We should feel comfortable with the concept of data now, but how about collecting, organizing, analyzing and interpreting the data? Here is another example of numbers:

4, 6, 7, 9, 18, 19, 28

To use these numbers in sports statistics, we need a context. (Remember, numbers + context = data.). To give the men their due, here we go:

2011 NEWMAC Cross Country Championships
October 30, 2011
Franklin Park, Boston MA

Event 1 Men 8k Run CC

		Name	Yr	School	Final	Pts
1	234	Dan Harper	SR	MIT	25:28	1
2	242	Roy Wedge	SO	MIT	25:34	2
3	239	Stephen Serene	SR	MIT	25:45	3
4	254	Ryan O'Connell	JR	Springfield	26:07	4
5	236	Ben Mattocks	SR	MIT	26:10	5
6	248	Brian Fuller	SR	Springfield	26:21	6
7	258	Anthony Salvucci	JR	Springfield	26:23	7
8	235	Allen Leung	FR	MIT	26:26	8
9	256	Zach Pietras	JR	Springfield	26:30	9
10	226	Trevor Siperek	SR	Coast Guard	26:33	10
11	204	Gary Ezzo	SO	Coast Guard	26:43	11
12	183	Mark Gulesian	FR	Babson	26:52	12
13	240	Logan Trimble	JR	MIT	26:56	13
14	206	Joe Hill	SR	Coast Guard	26:57	14
15	223	Kevin Obrien	SR	Coast Guard	26:59	15
16	232	Andrew Erickson	SR	MIT	27:00	16
17	238	Eric Safai	SO	MIT	27:03	
18	185	Nathan Kolman	SO	Babson	27:06	17
19	255	Matthew Peabody	SR	Springfield	27:10	18
20	241	Matt Weaver	SR	MIT	27:12	
21	246	David Birdsall	SR	Springfield	27:22	19
22	237	Jay McKenna	SO	MIT	27:25	

		Name	Yr	School	Final	Pts
23	276	Dominic Gonzalez	SO	WPI	27:31	20
24	231	Karl Baranov	FR	MIT	27:34	
25	233	Kris Frey	FR	MIT	27:34	
26	288	Andrew Zayac	FR	WPI	27:37	21
27	284	Lucas Smith-Horn	SR	WPI	27:40	22
28	271	Michael Richard	JR	Wheaton (MA)	27:41	23
29	188	Andrew Oram	SR	Babson	27:44	24
30	217	Myles McCarthy	SR	Coast Guard	27:46	25
31	214	Jordon Lee	SO	Coast Guard	27:49	26
32	283	Shane Ruddy	JR	WPI	27:50	27
33	259	Matthew Scully	JR	Springfield	27:51	28
34	192	Jake Williams	JR	Babson	27:52	29
35	225	Devin Quinn	SR	Coast Guard	27:54	30
36	221	Joseph O'Connell	FR	Coast Guard	27:54	
37	224	Bradley Pienta	SO	Coast Guard	28:02	
38	182	Brian deLeon	SO	Babson	28:05	31
39	274	Scott Burger	JR	WPI	28:10	32
40	190	AJ Skains	SO	Babson	28:11	33
41	222	Ryan O'Neil	SO	Coast Guard	28:17	
42	273	Josh Brodin	JR	WPI	28:18	34
43	219	Brian Musard	FR	Coast Guard	28:19	
44	194	Nathan Buck	SO	Clark (MA)	28:20	35
45	191	Mike Smith	SR	Babson	28:20	36
46	184	William Hallock	SO	Babson	28:24	
47	281	Brendan McKeogh	FR	WPI	28:26	37
48	181	Eduardo Arechabala	SO	Babson	28:36	
49	207	Stephen Horvath	FR	Coast Guard	28:40	
50	251	Corey Hamel	SO	Springfield	28:40	
51	263	Erich Voelker	SO	Springfield	28:41	
52	269	Karl Mader	JR	Wheaton (MA)	28:49	38
53	216	James Martin	FR	Coast Guard	28:56	
54	282	Hunter Putzke	SO	WPI	28:59	
55	250	James Gornell	SR	Springfield	29:00	
56	277	Rob Hollinger	SO	WPI	29:03	
57	198	Joe Kennelly	SR	Clark (MA)	29:04	39
58	189	Chris Pierce	JR	Babson	29:15	

		Name	Yr	School	Final	Pts
59	245	Marck Bashaw	FR	Springfield	29:16	
60	218	Matt Monahann	SO	Coast Guard	29:17	
61	275	Sean Dillon	SR	WPI	29:20	
62	280	Greg McConnell	SR	WPI	29:24	
63	278	Joseph Kelly	FR	WPI	29:26	
64	249	Joe Geurds	FR	Springfield	29:28	
65	287	Keegan Westwater	SO	WPI	29:31	
66	205	Thomas Heikkinen	JR	Coast Guard	29:33	
67	187	Leo Martinez	SR	Babson	29:35	
68	202	Alex Cropley	SR	Coast Guard	29:49	
69	272	Bradford Bailey	JR	WPI	29:53	
70	228	Drew Stafford	SO	Coast Guard	29:58	
71	197	Connor Joyce	SR	Clark (MA)	30:00	40
72	201	Andrew Cook	JR	Coast Guard	30:01	
73	195	Rob Gammel	SO	Clark (MA)	30:03	41
74	252	Tyler Leahy	FR	Springfield	30:05	
75	213	Jon Lash	FR	Coast Guard	30:15	
76	247	Scott Bushey	FR	Springfield	30:19	
77	268	John Green	FR	Wheaton (MA)	30:21	42
78	285	Michael Szkutak	SR	WPI	30:23	
79	286	Mark Vanacore	SR	WPI	30:25	
80	203	Patrick Dreiss	FR	Coast Guard	30:26	
81	220	Bradley Nelson	FR	Coast Guard	30:28	
82	193	Scott Worth	JR	Babson	30:32	
83	212	Zachary Kearney	FR	Coast Guard	30:33	
84	265	Sean Astle	SR	Wheaton (MA)	30:36	43
85	260	Mike Sherry	FR	Springfield	30:38	
86	229	John Tabb	SO	Coast Guard	30:45	
87	266	Harry Bachrach	FR	Wheaton (MA)	30:56	44
88	209	Erick Jackson	FR	Coast Guard	30:57	
89	200	Kyle Sullivan	FR	Clark (MA)	31:06	45
90	208	Ryan Hub	FR	Coast Guard	31:43	
91	186	Michael Liachowitz	SO	Babson	32:28	
92	261	Dan Sugar	FR	Springfield	32:35	
93	227	Zachary Speck	JR	Coast Guard	32:40	
94	199	Sam Morrison	JR	Clark (MA)	33:18	46

		Name	Yr	School	Final	Pts
95	270	Christopher Panzini	SO	Wheaton (MA)	34:00	47
96	264	Nathan Watson	FR	Springfield	35:00	
97	262	Robert Sullivan	FR	Springfield	35:17	
98	267	James Court	SO	Wheaton (MA)	36:00	48
99	253	Christopher Malia	SR	Springfield	37:10	
100	196	Thomas Hurlburt	SR	Clark (MA)	38:20	49

First COLLECTING ...

The numbers above correspond to the scoring places of the Springfield College men's cross country team at the 2011 NEWMAC championships. Here is how the data was **COLLECTED**:

The NEWMACs were held at Franklin Park in Boston and hosted by MIT. All seven men's teams from the NEWMAC ran their varsity and junior varsity runners together in one race. When each runner crossed the finish line, he was given a Popsicle stick with a number on it. The number corresponded to his place. (The first runner to finish got number 1, the second runner number 2, etc.) The runners then went to the scorer's table and an official **COLLECTED** all the sticks, noting the runner's name and school. When all the runners on all seven teams had turned in their sticks, the data **COLLECTION** phase was complete. The official now had the list of finishers above.

Then ORGANIZING ...

Now the **ORGANIZATION** phase began. The official's job was to certify who had won the conference championship. He took all the runners and organized them by school. He **ORGANIZED** the data by grouping all the runners with their teams.

Then ANALYZING ...

The **ANALYSIS** portion involved adding up the team scores. That was accomplished by adding together the top five places for each team.

Finally INTERPRETING ...

The officials could **INTERPRET** the scores by looking for the lowest team score. The team with the fewest points is declared the winner. And the team with the second fewest points is the runner-up. The order of finish is listed below:

Team Scores

	Team		Total	1	2	3	4	5	6	7
1	**MIT**		19	1	2	3	5	8	13	16
	Total Time:	2:09:23.00								
	Average:	25:52.60								
2	**Springfield**		44	4	6	7	9	18	19	28
	Total Time:	2:12:31.00								
	Average:	26:30.20								
3	**Coast Guard**		75	10	11	14	15	25	26	30
	Total Time:	2:14:58.00								
	Average:	26:59.60								
4	**Babson**		113	12	17	24	29	31	33	36
	Total Time:	2:17:39.00								
	Average:	27:31.80								
5	**WPI**		122	20	21	22	27	32	34	37
	Total Time:	2:18:48.00								
	Average:	27:45.60								
6	**Wheaton (MA)**		190	23	38	42	43	44	47	48
	Total Time:	2:28:23.00								
	Average:	29:40.60								
7	**Clark (MA)**		200	35	39	40	41	45	46	49
	Total Time:	2:28:33.00								
	Average:	29:42.60								

In terms of times, SC runners averaged 26:30.20 per runner on Franklin Park's 8K course. We finished second to MIT. MIT were the men's NEWMAC champions, and their runners averaged 25:52.60. Even though MIT dominated the meet, SC was only 38 seconds per runner behind. If each returning SC runner bettered their times by only 39 seconds, we could beat MIT! Over 8K, 39 seconds is very doable. But will our runners do the work in track (increased speed) and next summer (strength and endurance) to make those 39 seconds happen? Or will some other team step up?

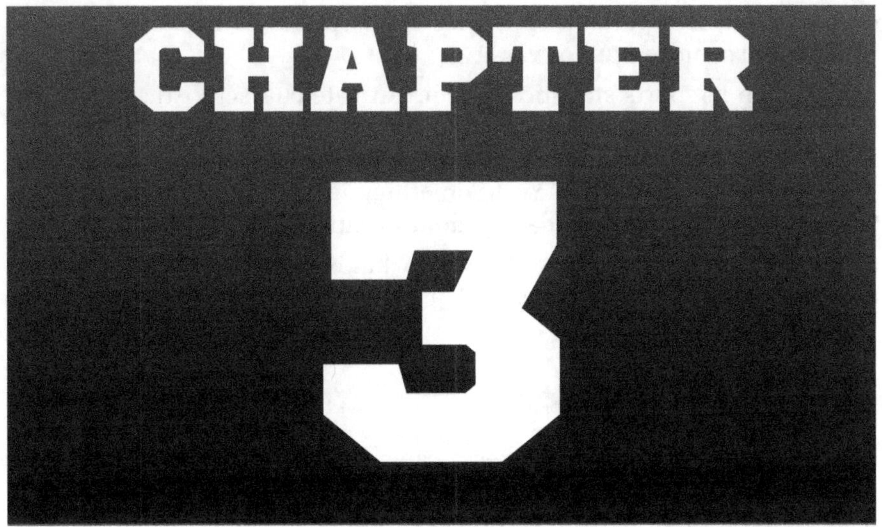

CHAPTER 3

Variable variables

If you dwell on statistics, you get shortsighted. If you aim for consistency, the numbers will be there at the end.

— Tom Seaver, Hall of Fame pitcher

PREVIEW: THE DIFFERENCE BETWEEN CATEGORIES AND NUMBERS

What is a variable?

When we review the data we have seen so far — that is the women's and men's cross country results from the last chapter — we notice that all the finishing places are different. They vary from one runner to the next. That means they are **variable**. The finishing place for each woman is a variable, because it varied from runner to runner. Their names are variables. Their colleges are variables. Their times are also a variable. Now two runners could have had the exact same time (unlikely, but they could have), in that case is it a variable? Yes, because time can vary from runner to runner.

How about on the men's side? The varsity and junior varsity ran in the same race. Are varsity and junior varsity different? Yes! Even though they ran in the same race, the runner's level of competition, or classification, is a variable. It varies from runner to runner; some are varsity, and some are junior varsity.

Our job in sports statistics is to investigate characteristics that vary. More importantly we will ask, "What can the variation tell us? Does the variation *mean* something? Is it *significant*?" We will run tests on the data. The tests will tell us, "Yes! Something important is happening!" Or "No, it's just ordinary everyday random variation."

Luckily for us, there are only two kinds of variables in statistics: categorical and quantitative. Let us begin by understanding both variables.

Categorical variable

A **categorical** variable describes a category, or a group. If what varies are categories, it is a categorical variable. Like position on a basketball team (guard, forward, center), unit on a football team (offense, defense, special teams), country of origin of a hockey team (USA, Canada, Finland). Here are several other examples of categorical variables.

Categorical variable	Levels or group
Level of competition	Varsity
	Junior varsity
	Freshman
	Club
	Intramural
Class	Freshman
	Sophomore
	Junior
	Senior
	Graduate student
Academic level	Academic All-American
	Not Academic All-American status

OK, sounds good. Any variable that can be broken down into groups or categories is a categorical variable. There are only two types of variables, and the first seems pretty easy. How about the second one?

Quantitative variable

Well, the other type of variable is a **quantitative** variable. And the hint here is the root of the word quantitative: quantity. Yes, a quantitative variable deals with quantities, or numbers. If what varies are numbers, it is a quantitative variable. Like the runners times. 26:32.54 is quantitative. It is a number. The height of each runner is quantitative. The weight of each runner is quantitative. But varsity is a group. It is categorical. Make sense?

Categorical or quantitative?

Here are some characteristics from football that vary from player to player, so they are variables. Think about each one and decide as you read them whether they are categorical variables or quantitative variables.

- Position
- Height
- Weight
- Team (offense, defense or special teams)
- 40 time
- Division (I, II or III)
- Number of tackles
- Points scored each game

Let's see how you did. Positions include linebacker, running back, kicker, etc., and they are all categories or groups (no numbers!) so position is a categorical variable. Height (6', 6'2", etc.) are all numbers, they are quantities, which means height is a quantitative variable. Weight (160 pounds, 190, 225, etc.) is represented by a number, so weight is a quantitative variable. Team (offense, defense or special teams) is a categorical variable. 40 time (4.25, 5.1, 5.99, etc.) is a quantitative variable. Division (I, II or III) is a categorical variable. This one is interesting because I, II and III are numbers, but in this case they are being used to describe a category or group. For instance Division III is a group of colleges that share the same commitment to the student-athlete. Number of tackles (2, 35, 112, etc.) is a quantitative variable (the description NUMBER of tackles, tells us right off the bat). And finally points scored per game (40, 41, 55, etc.) is a quantitative variable.

Here's a tricky one. Every column below is a variable. Rank is a column. Rank is a variable. Is rank categorical or quantitative? Those are the only two choices.

Football AP Poll Jan 14 2020

Rank	Team	Record	PTS	1st	Prev	Chg
1	LSU	15-0	1550	62	1	—
2	CLEM	14-1	1487	0	3	1
3	OSU	13-1	1426	0	2	-1
4	UGA	12-2	1336	0	5	1
5	ORE	12-2	1249	0	7	2
6	FLA	11-2	1211	0	6	—
7	OKLA	12-2	1179	0	4	-3
8	ALA	11-2	1159	0	9	1
9	PSU	11-2	1038	0	13	4
10	MINN	11-2	952	0	16	6
11	WIS	10-4	883	0	11	—
12	ND	11-2	879	0	14	2
13	BAY	11-3	827	0	8	-5
14	AUB	9-4	726	0	9	-5
15	IOWA	10-3	699	0	19	4
16	UTAH	11-3	543	0	12	-4
17	MEM	12-2	528	0	15	-2
18	MICH	9-4	468	0	17	-1
19	APP	13-1	466	0	20	1
20	NAVY	11-2	415	0	21	1
21	CIN	11-3	343	0	23	2
22	AFA	11-2	209	0	24	2
23	BSU	12-2	188	0	18	-5
24	UCF	10-3	78	0	0	—
25	TEX	8-5	69	0	0	—

Rank is a number, so it is numerical, right? And numerical is quantitative, a quantity. But does average make sense?

I add them all together ($1 + 2 + 3 + ...+ 24, + 25 = 325$) then divide by the total of 25 teams.

$$\frac{325}{25} = 13$$

Each of the teams has an average ranking of 13?

No, average does not make sense. LSU may have been the best college football team of all time, and they have an average ranking of 13? No again!

If average does not make sense, the variable is categorical.

LSU's average ranking of 13 does not make sense, but LSU is ranked number 1. It has a grouping, a category. It has just LSU in the group so it is a boring category, but a category it is!

PRACTICE PROBLEM

Here are 9 variables. (Each column is a variable.) List them and identify whether they are categorical or quantitative.

Tour de France Winners (1999-2019)

Year	Winner	Country	Time	Avg speed	Stages	Dist (km)	Starters	Finishers
1999	Lance Armstrong*	USA	91.32.16	40.30	20	3687	180	141
2000	Lance Armstrong*	USA	92.33.08	39.56	21	3662	180	128
2001	Lance Armstrong*	USA	86.17.28	40.02	20	3453	189	144
2002	Lance Armstrong*	USA	82.05.12	39.93	20	3278	189	153
2003	Lance Armstrong*	USA	83.41.12	40.94	20	3427	189	147
2004	Lance Armstrong*	USA	83.36.02	40.53	20	3391	188	147
2005	Lance Armstrong*	USA	86.15.02	41.65	21	3608	189	155
2006	Oscar Periero	SPA	89.40.27	40.78	20	3657	176	139
2007	Alberto Cantadar	SPA	91.00.26	38.97	20	3547	189	141
2008	Carlos Sastre	SPA	85.52.22	40.50	21	3559	179	145
2009	Alberto Contador	SPA	85.48.35	40.31	21	3460	180	156
2010	Andy Schleck	LUX	91.59.27	39.59	20	3642	195	167
2011	Cadel Evans	AUS	86.12.22	39.79	21	3430	198	167
2012	Bradley Wiggins	GBR	87.34.47	39.83	20	3488	198	153
2013	Chris Froome	GBR	83.56.20	40.55	21	3404	198	169
2014	Vicenzo Nibali	ITA	89.59.06	40.69	21	3661	198	164
2015	Chris Froome	GBR	84.46.14	39.64	21	3360	198	160
2016	Chris Froome	GBR	89.04.48	39.62	21	3529	198	174
2017	Chris Froome	GBR	86.20.55	41.00	21	3540	198	167
2018	Geraint Thomas	GBR	83.17.13	40.23	21	3349	176	145
2019	Egan Bernal	COL	82.57.00	40.58	21	3349	176	155

* stripped due to doping

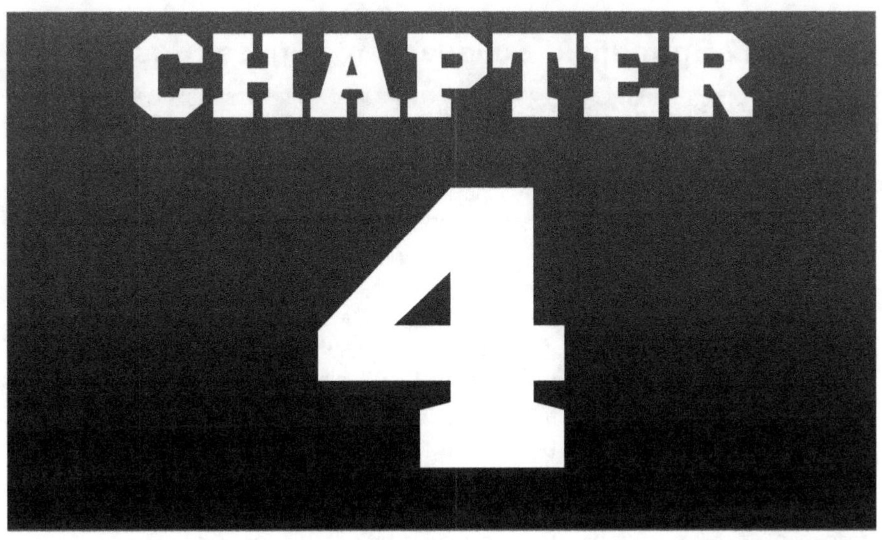

A chart is worth a 1000 pieces of data

When I negotiated Bob Stanley's contract with the Red Sox, we had statistics demonstrating he was the third-best pitcher in the league. They had a chart showing he was the sixtieth-best pitcher on the Red Sox!

— Bob Woolf, player agent

PREVIEW: BAR CHART OR PIE CHART? DECISIONS, DECISIONS

Now that we're comfortable with which variables are categorical and which are quantitative, we can turn to how we can best understand them. How can we explain them to others. Well, the best way is with a picture. After all, a picture is worth a thousand words. And with categorical variables, the two best pictures are a **bar chart** and a **pie chart**.

To create a chart we need data. Let's **COLLECT** some data. (Remember, our definition of statistics was to collect, organize, analyze and interpret data.) Here is the 2012 Springfield College men's soccer team from the Springfield College website. I've sorted it alphabetically by class to make it easier to use.

2012 Springfield College Men's Soccer Roster

	Name	Class	Pos	Ht	Hometown	Major
0	Billy Schmid	Fr	GK	5-10	Wethersfield, CT	Physical Education
23	Marco Callisto	Fr	M	5-6	Mamaroneck, NY	Undeclared
24	Ben Marcus	Fr	M	5-8	Wappingers Falls	Athletic Training
25	Ryan McAdoo	Fr	M	5-7	Terryville, CT	Physical Education
26	Austin Vico	Fr	F	5-10	Plymouth, MA	Criminal Justice
27	Nick Kobel	Fr	M	6-1	Webster, MA	Sport Management
28	Alejandro Miguel	Fr	D	5-9	Bellmore, NY	Physical Education
6	Danny Amato	Jr	M	5-8	Newington, CT	Physical Education
8	Scott Morneault	Jr	M	5-9	Bristol, CT	Physical Education
11	Collin Smith	Jr	D	5-11	Byfield, MA	Elementary Education
13	John Mankus	Jr	D	6-0	Vernon, CT	Business Administration
14	Michael Fowler	Jr	F	5-9	Maynard, MA	Sport Management
16	Scott Saucier	Jr	M	5-9	Derry, NH	Sport Management
19	Drew Vanasse	Jr	M	6-1	Concord, MA	Sport Management
22	Ryan Malone	Jr	M	6-2	Chicopee, MA	Sport Management
77	Mike Breault	Jr	GK	5-10	Seymour, CT	Physical Education
1	Brett Bascom	So	GK	6-2	New London, NH	Recreation Management
3	Tyler Allen	So	M	5-7	Southbridge, MA	Biology
4	Mike Desroches	So	F	5-4	Enfield, CT	Physical Education
5	Joe LaBella	So	M	5-6	Wethersfield, CT	Recreation Management
7	Dan O'Grady	So	D	5-8	Somers, CT	Physician Assistant
9	Zach Dutter	So	M	5-8	Dallas, PA	Recreation Management
15	Kevin Nowak	So	M	6-0	Plainsboro, NJ	Applied Exercise Science
17	Logan Murphy	So	F	6-2	Troy, NY	Sport Management
18	Drew Sommer	So	D	5-10	West Chester, PA	Physical Education
21	Brian Dunn	So	D	5-8	Fiskdale, MA	Physical Education
2	Tyler Fletcher	Sr	M	5-11	Middletown, CT	Sport Management
10	Alex Reilly	Sr	M	5-10	Ledyard, CT	Physical Education
12	Joseph McSpiritt	Sr	D	6-0	Mill River, MA	Physical Education
20	David Chessen	Sr	D	6-0	Edison, NJ	Physical Therapy
74	Chris Walton	Sr	GK	6-5	Cromwell, CT	Applied Exercise Science

We want to use this data to answer the question,

Describe our men's soccer team by class.

We have four classes, typically we would expect those classes to be equally spread out on the roster. Four class's means ¼ of the athletes are freshman, ¼ sophomores, ¼ juniors, and ¼ seniors. Those are the expected values, ¼ or 25% in each class. To answer our question of describing the team by class, we will analyze the actual numbers (observed numbers) and compare them to our expected values of 25%.

If we count how many are freshman, sophomore, junior and senior, that will answer the question for us. The count is the observed frequency and the percentage is the relative frequency. Our next step is to ORGANIZE the data. We will organize it in a frequency table broken down by class (freshman, sophomore, junior and senior). The table on the next page shows how frequently each class shows up on the team, so we call it a **frequency table**.

The frequency row summarizes how many players on the team are freshman, sophomore, etc. I got that from the roster. I simply counted how many players were in each class. I then organized it into a table so it would be easier for me to use. The number of players in each class we will call the frequency. The percentage of players in each class we call the relative frequency.

Here is our frequency table:

Springfield College Men's Soccer

	Freshman	Sophomore	Junior	Senior	Total
Frequency	7	10	9	5	31
Relative frequency	22.6%	32.3%	29.0%	16.1%	100%

Next we **ANALYZE**. Let's find what percent each class is of the total. We need to know the total number of players. I counted all the players on the roster. There were 31 men playing soccer for Coach Siebert in 2012. First, let's just look at the freshmen. There were 7 of them, 7 of the 31 total players were freshmen. Then I divided 7/31 to get .226. That is the **proportion** that are freshmen. Move the decimal point two places over to the right and we get the **percent** who are freshmen, 22.6%. It's also called the **relative frequency**. That just means how the frequency *relates* to the total.

We have now collected and organized the data. We analyzed it by calculating the relative frequencies. We know what percent of the team is in each class. Finally, we get to **INTERPRET** the data. The SC men's soccer team is predominantly from the two middle classes in 2012. 32.3% are sophomores and 29% are juniors, so 61.3% are in the middle classes (as opposed to 38.7% who are freshmen and seniors). That means they are an inexperienced team in terms of senior leadership, but an experienced team in terms of players who are comfortable in Coach Siebert's system.

That is it! We just completed our first exercise in sports statistics. We collected, organized, analyzed, and interpreted data on the men's soccer team. Was this an important question? Yes, it might be! If you are a recruit looking for a Division III school where you can get some playing time as a freshman, the small number of players who are seniors means only five players will graduate. That does not create a lot of openings.

How about if you were the coach? You would need to coach this team differently because the freshmen may get minimal playing time. Also, the senior leadership gap needs to be filled, the coach and assistants need to step in. Maybe the team needs different practices to work on the fundamentals of the offense rather than on specific plays which might enhance a weakness on the opposing team. Remember, the freshmen are still learning the coach's system.

The athletic director might have a question. 31 players means about 8 players per class (31 divided by the 4 classes is 7.75 or about 8), but there are only 5 seniors. What is it about the team that has the players not returning for their senior year? Or maybe it is not the team. Maybe it is a college-wide phenomena. Do all teams have relative frequencies similar to men's soccer?

Bars and pies: picturing data in Minitab

But let's return to our data. Our original goal was to graph our data in two ways, in a bar chart and a pie chart. While the data we're using is pretty basic, pictures can make it easier to compare the numbers.

We will make our first graph on Minitab. Let's graph the men's soccer team. Click on the start icon in the bottom left of your computer screen, then click all programs and pick Minitab. Or type Minitab into the search box and select it. Do not use Minitab express.

Enter data in Minitab

In the box under C1 type Class. In the box under C2 type Frequency. In the boxes under Class type Freshman, Sophomore, Junior, and Senior. Then in boxes 1-4 under Frequency type 7, 10, 9, 5.

Choose Minitab options

Select Graph from the menu bar. Choose **Bar Chart**. From the drop down menu pick Values from a table. That is because you just typed in your values in C1, which is the "table" Minitab is going to use. Keep Simple Graph highlighted and click OK. Then click in the box under Graph Variables once to move your curser to the box, and that will bring up Frequency and Class. Scroll over and double click on Frequency. That moves Frequency into the Graph Variable box. Then click on the box to the right of Categorical Variables and scroll over to the left and double click on Class. You will see Class move to the Categorical Variable box (appropriate because class is a categorical variable).

Select Labels and type in 2012 Springfield College Men's Soccer. Then subtitle 1 type in Roster by Class. Subtitle 2 is your name. Click OK.

In the footnote you will describe our men's soccer team by class. It is a predominantly filled with sophomores and juniors. That is your interpretation which will always go in the footnote.

For a **Pie Chart**, click on Pie Chart and then select Chart values from a table. Click in the box labeled Categorical Variables and then double click on C1 on the left. Then click in box labeled Summary Variables and double click on C2. Then click on Labels, and for the Title type in Springfield College Men's Soccer 2012. Under subtitle 1 type Roster by Class and under subtitle 2 type in your name.

Done!

Here is a picture from Minitab, including the inputs in C1 and C2, and our output (the two graphs):

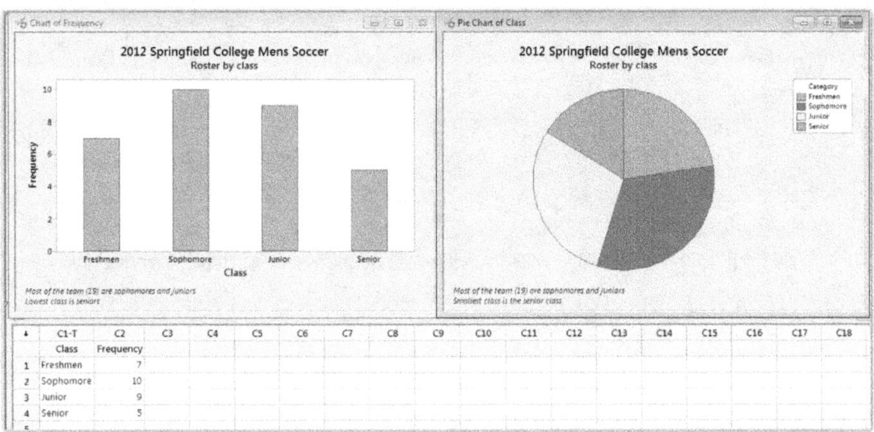

	C1-T	C2	C3	C4	C5	C6	C7	C8	C9	C10	C11	C12	C13	C14	C15	C16	C17	C18
	Class	Frequency																
1	Freshmen	7																
2	Sophomore	10																
3	Junior	9																
4	Senior	5																

Note: Bar charts and pie charts display the same data, just in different ways. They are interchangeable. Which to use is totally your choice. On a test either one would be right.

To decide which to use, you can ask yourself which makes it easier to see the point you're making with your data. When designing a graph, you have extraordinary power to sway your reader's opinion. People can do most anything with data, twisting it to support their argument. Imagine when an agent goes into a contract negotiation with a general manager. Both will be prepared with binders and binders full of statistics to support their argument for more money (the agent) or less money (the general manager). But there was only one athlete and one set of data, right? Yes, but both sides are doing their utmost to twist the data to support their side of the argument.

Now, let's answer the athletic director's question. Is the low number of seniors in soccer a men's problem or does it affect the women too? We'll look at the 2012 Springfield College women's soccer team, sorted by class.

2012 Springfield College Women's Soccer Roster

	Name	Class	Pos	Ht	Hometown	Major
4	Nicole Fowler	Fr	F	5-5	Maynard, MA	Physical Therapy
8	Brooke Hattinger	Fr	M	5-5	Weymouth, MA	Biology
12	Dana Hebert	Fr	D	5-3	Kensington, CT	Physician Assistant
16	Krissy Cicalis	Fr	M	5-2	Bridgewater, MA	Physician Assistant
18	Heather Slavin	Fr	M	5-4	Enfield, CT	Physician Assistant
26	Carly Mappa	Fr	F	5-3	Fairfield, CT	Sport Management
2	Karly Barrett	Jr	M	5-6	Sparta, NJ	Occupational Therapy
5	Ashley Carressi	Jr	M	5-0	Billerica, MA	Elementary Education
6	Brittany Samson	Jr	D	5-5	Chicopee, MA	Sport Management
11	Kim Rasmussen	Jr	F	5-6	East Berlin, CT	Physical Education
17	Nina Vital	Jr	F	5-3	Ludlow, MA	Physician Assistant
20	Vicky DiNatale	Jr	D	5-7	Weymouth, MA	Appl Exercise Science
21	Karleigh Bradbury	Jr	M	5-8	South Portland, ME	Appl Exercise Science
24	Breena Salwocki	Jr	M	5-4	Newington, CT	Physical Education
0	Holly Ouellette	So	GK	5-3	Shrewsbury, MA	Physical Education
3	Alicia Leo	So	D	5-5	West Springfield, MA	Health Science
23	Kelly Haines	So	D	5-9	Easton, CT	Physical Education
25	Kelly Ziogas	So	D	5-5	Bristol, CT	Physical Therapy
0	Erin Greenstein	Sr	GK	5-3	Framingham, MA	Occupational Therapy
9	Sara Dalton	Sr	M	5-5	Schenectady, NY	Athletic Training
10	Jordana Harrison	Sr	M	5-5	Saco, ME	Health Administration
13	Courtney Letourneau	Sr	M	5-3	Paxton, MA	Sports Biology
14	Lauren Muser	Sr	D	5-5	Derry, NH	Sports Biology
15	Erica Donnelly	Sr	F	5-6	Weymouth, MA	Health Science
19	Mary Hogan	Sr	D	5-5	Farmington, CT	Physical Therapy
22	Britt Edwards	Sr	D	5-9	West Springfield, MA	Elementary Education

Make a frequency table (remember to include both the frequency and the relative frequency) for the women.

Springfield College Women's Soccer

	Freshman	Sophomore	Junior	Senior	Total
Frequency	6	4	8	8	26
Relative frequency	23.1%	15.4%	30.8%	30.8%	100%

When you enter the data into Minitab it will look like this:

	C1	C2	C3
	Class	Frequency	
1	Freshman	6	
2	Sophomore	4	
3	Junior	8	
4	Senior	8	

Then we can have Minitab produce a bar chart and a pie chart for the women:

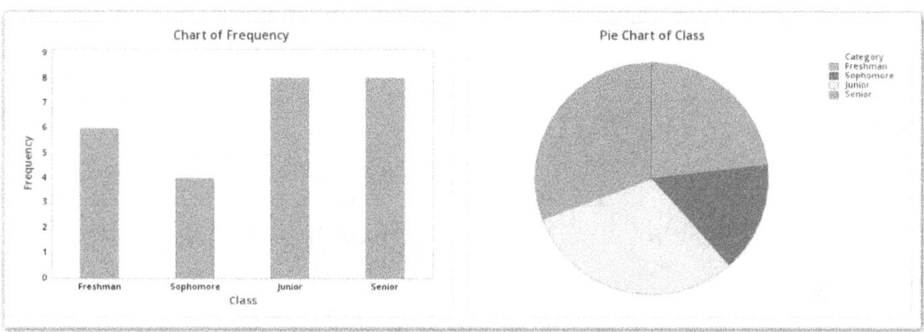

Interesting, the men have 31 roster spots while the women had 26 in 2012. The men have more sophomores and juniors, while the women have more juniors and seniors. Seniors are not a problem for the women's team. It would be of interest to see if that is still true today.

The reason we looked at the women's data was to see if the low number of seniors was a men's team problem or something more indicative of the athletics program. Well, we see that it is a men's soccer problem, not a women's soccer problem.

If we wanted to look further into the senior drop off we could check the current rosters of other teams. Maybe with the boys playing in soccer leagues from the age of four they are burned out by the time they are 21.

Perhaps they decide to pursue other activities, like graduating and finding a job.

One more point about bar charts. Categories can go in any order you want. Here are the average salaries in professional sports. The first bar chart lists the sports in the order of my most favorite to least favorite sports. We can see salaries increase as my interest decreases. (Except for soccer!) The second bar chart lists the salaries from highest to lowest. When we order them from highest to lowest we call it a Pareto chart.

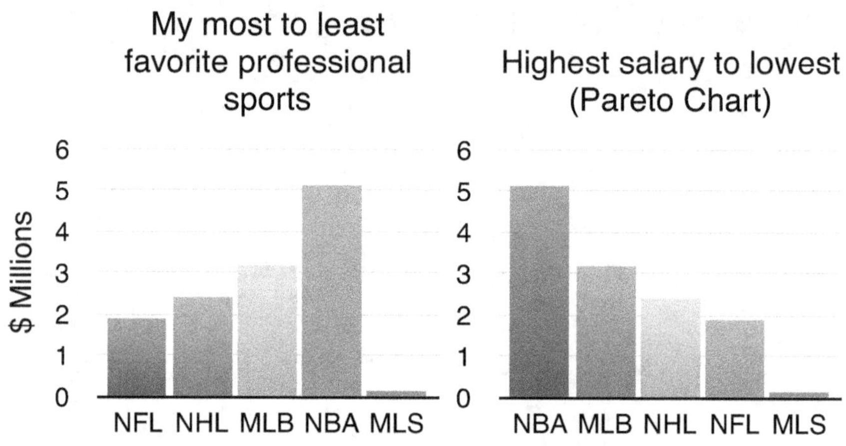

PRACTICE PROBLEMS

Following is the 2014 roster for Boston's pro-lacrosse team, the Boston Cannons. Create an appropriate graph by position.

2014 Boston Cannons Roster

	Name	Position	Height	Weight
20	Owen Blye	Attack	6' 3" (191 cm)	185 lbs.
27	Kevin Buchanan	Attack	5' 11" (180 cm)	180 lbs.
21	Jim Connolly	Attack	5' 11" (180 cm)	175 lbs.
1	Will Manny	Attack	5' 9" (175 cm)	160 lbs.
85	Mitch Belisle	Defense	5' 10" (178 cm)	195 lbs.
55	Eric Martin	Defense	6' 3" (191 cm)	220 lbs.
72	Scott McWilliams	Defense	6' 3" (191 cm)	205 lbs.
17	Brodie Merrill	Defense	6' 4" (193 cm)	205 lbs.
43	Jack Reilly	Defense	6' 3" (191 cm)	215 lbs.

	Name	Position	Height	Weight
77	Kyle Sweeney	Defense	6' 2" (188 cm)	195 lbs.
5	Jordan Burke	Goalie	6' 1" (185 cm)	190 lbs.
19	Austin Kaut	Goalie	6' 1" (185 cm)	197 lbs.
28	Brent Adams	Midfield	6' 1" (185 cm)	165 lbs.
12	Martin Bowes	Midfield	6' 2" (188 cm)	200 lbs.
22	Craig Bunker	Midfield	5' 9" (175 cm)	185 lbs.
24	Chris Eck	Midfield	6' 0" (183 cm)	220 lbs.
42	Rob Emery	Midfield	6' 3" (191 cm)	210 lbs.
3	Brendan Porter	Midfield	6' 4" (193 cm)	215 lbs.
99	Paul Rabil	Midfield	6' 3" (191 cm)	220 lbs.
2	Scott Ratliff	Midfield	6' 0" (183 cm)	185 lbs.
11	Matt Smalley	Midfield	5' 10" (178 cm)	180 lbs.
(na)	Erik Smith	Midfield	5' 9" (175 cm)	185 lbs.
41	Mike Stone	Midfield	5' 11" (180 cm)	175 lbs.

Pick your favorite team of any sport and create a graph of the roster by a category such as position or class (for college).

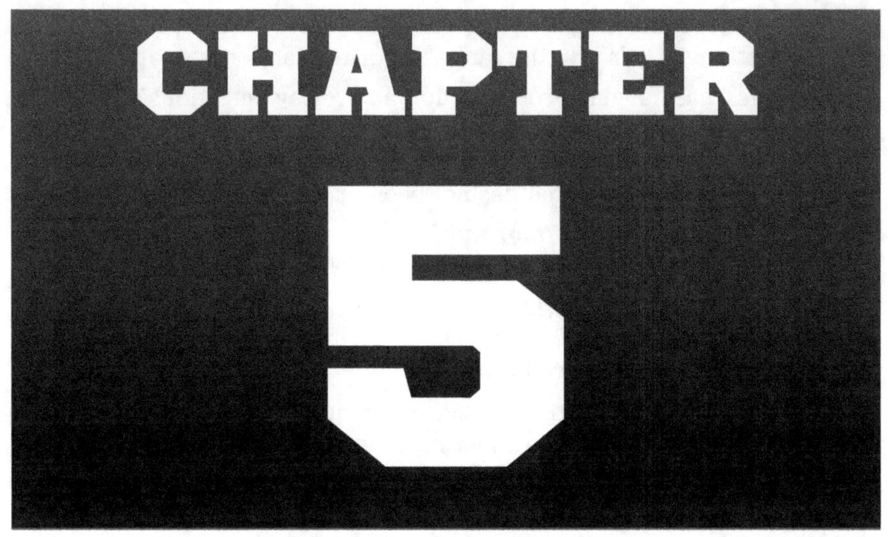

Time out for testing

Who says there's an unemployment problem in this country? Just take the five percent unemployed and give them a baseball stat to follow.

— *Andy Van Slyke, Pirate outfielder*

PREVIEW: YOUR FIRST STATISTICAL TEST

Now that we've had some experience looking at data, asking questions, finding answers and interpreting those answers, let's take a time out to look at why we're doing this.

Numbers are everywhere. Tom Brady threw 4 interceptions. Patrice Bergeron scored 2 goals. Tuukka Rask made 31 saves.

When the numbers are unusual and unexpected we ask questions. What do they mean? How will Tom Brady perform in Tampa? Is David Pastrnak the best scorer on the Bruins? Is Tuukka Rask ready to lead the Bruins to a Stanley Cup? Are the Celtics better with Kemba vs. Kyrie?

Sports talk radio is kept busy by people who are certain their opinions — their guesses — about what the numbers mean are just as good as anyone else's.

But to make solid decisions — and sound knowledgeable on sports radio — we need more than gut feelings. We want a scientific way to tell us whether the numbers are significant. Does a big jump in interceptions mean something? Or is it in the range of normal random variation?

Statistics has formal processes for answering questions like these. One is called a **hypothesis test**. A hypothesis test is just like it sounds. It tests a hypothesis. A hypothesis is a guess based on specific observations. It is a guess that has not been proven or disproven yet.

Is a roster with 5, 5, 5, 0 for the classes a problem? Are there issues with this team? Is there something going on with this team?

Is a roster with 25, 25, 26 27 just random variation?   That would mean there are no issues for this team. There is nothing going on with this team if it is random variation.

How can we tell if there is something going on with a team's roster, or if the differences we see are just random variation? We can test the data to see how well it fits our expectations.

Introducing a chi-square goodness of fit test that will do just that. We couldn't answer that in Chapter 4. With a hypothesis test we can.

Six steps to a hypothesis test

Our hypothesis tests throughout the course will follow the same six steps.

1. **Null hypothesis**
    Our first step is to ask, "What do we expect the data to look like normally?" What would it look like if there were nothing influencing it? In the soccer roster we looked at, we would expect teams to draw equally from each class, that is 25% freshman, 25% sophomores and so on. We know it won't be this exactly. It will have some natural random variation. But it will be close.

    We call this hypothesis our **null hypothesis** and label it as H_0, h-zero. Think of it as null for zero change. Null equals no. No change. No difference. Nothing going on. Null hypothesis. That is the hypothesis we test.

    The null hypothesis for the soccer data would be that the team is equally spread between all the classes, that is:

    H_0: Freshmen = Sophomores = Juniors = Seniors.

2. **Alternative hypothesis**
    Our second step is to say something is disturbing our data. Something is influencing it. The differences from the ideal aren't just ordinary random variation. We call this our **alternative**

hypothesis and label it H_A.

The alternative hypothesis for the soccer data would be that the team is *not* equally spread between all the classes, that is:

H_A: Freshmen ≠ Sophomores ≠ Juniors ≠ Seniors.

The two hypotheses are mutually exclusive. They can't both be true.

Isn't that redundant? Why have two hypotheses when one is just the opposite of the other? That's a good question! But once we dig into analyzing and interpreting results you will see how helpful it is when we're very clear on what we expect the data to look like ideally and what it looks like when it's not behaving normally.

Here's another head-scratching statistics idea. When we run a test, the result will either reject the idea there's no outside influence — reject the null hypothesis — or fail to reject that there's no outside influence — fail to reject the null hypothesis.

Whoa. What's this fail to reject stuff? Why not just say accept? Because they aren't the same thing! Accept means the null hypothesis is true, that the variation is with absolute certainty random. But we can't ever be 100% certain that the variation is plain, ordinary garden-variety variation. When we fail to reject the null hypothesis we're saying we don't have enough evidence to say the variation is out of the ordinary. It might be normal variation. It might not be.

In the soccer example, we will either reject that variation in classes is random or fail to reject that the variation in classes is random.

It's okay if that's not clear. Just let it sit. It will get clearer the more we work with real data.

3. **Test statistic**
   The third step is the test statistic. Through the magical powers of statistical formulas we can transform an entire set of data into ... a single number! We call it the test statistic. It's a numerical summary of the data. The data can have 5 data points or a million and it can be reduced to just one number. The value of the test statistic will tell us how strong our evidence is in our analysis.

   Back in the Dark Ages of my own college days we needed to tediously type all the data into a calculator and crunch the numbers ourselves.

Now we can load the whole data set into Minitab, tell it what formula to use and with a press of a button get a test statistic. In this course we will use four test statistics: χ^2 (chi-square, pronounced kigh, like high), z, t, and f. But our first test will use chi-square.

We will calculate the different-test statistics depending on what we know about the data and what question we're asking. Chi-square tests are good for categorical data. In the next chapter, we'll work with the soccer team data and use the chi-square test statistic, again.

4. ***p*-value**
   Next we calculate the *p*-value. This is the probability that the variation is not random. A lower probability means our variation is less likely to be random.

   For instance, how probable is it that we could expect 7 or 8 of each class on a team of 31 and end up with a team of 7, 10, 9, 5 without something affecting the students' decision to join or leave the team?

5. **Your statistical conclusion**
   What do you conclude from the test statistic and *p*-value? Do you reject the null hypothesis or fail to reject it? Between the *p*-value and the test statistic, the *p*-value is easier to interpret, so in Step 5 we will use the *p*-value for our conclusion. There are two possible:

   • Since the *p*-value is greater than the alpha level, I fail to reject H_o.
   • Since the *p*-value is less than the alpha level, I reject H_o.

   For sports stats, we will use an alpha level of .05 for all problems. That .05 is also known as our significance level.

6. **Your interpretation**
   This is where you become a sports statistician. You will answer what does it mean? If the *p*-value is < .05, there is significance. Something is going on. If *p*-value is > .05, there is nothing going on. The differences we see are just random variation.

Here are the steps all together. You'll see this list many times throughout the book.

Hypothesis Test

1. H_0:
2. H_A:
3. **Test statistic and its value:**
4. ***p*-value:**
5. **Statistical conclusion:**
6. **Interpretation:**

Don't worry if it's still muddy. It will become much clearer as we work with real data and ask real questions, beginning with examples in Chapter 6.

CHAPTER 6

Fit fest stats style

> Some say you have to use your five best players, but I found out you win with the five that fit together best as a team
>
> — *Red Auerbach, Hall of Fame basketball coach*

PREVIEW: GOODNESS-OF-FIT TEST TESTS HOW GOOD THE FIT IS

Now let's go back to the soccer rosters and look at the data for each class. Since there are four classes, we would expect each class to average 25% of the roster, right? For example, if you had a team with 100 players, you might expect 25 freshmen, 25 sophomores, 25 juniors, and 25 seniors. With all else being equal, each class would likely include 25% of the team. We saw in the previous chapter this isn't the case with men's soccer at Springfield College. We saw that the soccer teams are not equally divided among the four classes.

Is there an issue with soccer at Springfield College or is it just plain randomness? Class is a variable, right? By definition, it varies. In our above example, if we had 100 players, would a distribution of 24, 26, 25 and 25 be unlikely? No. Sometimes you have better recruiting classes, or more injuries, or academic issues, disciplinary issues, etc. There is almost always a randomness to the distribution. We might be surprised to see 25, 25, 25, and 25 exactly!

Our question for this chapter becomes: Is the class distribution of student-athletes on the Springfield College soccer teams **significantly** different from the expected frequencies of 25% or are the differences we see due to ordinary random variation?

To answer that question we will use the **chi-square goodness-of-fit test**. As was mentioned, this test is good when we have categorical data and want to know how well it fits our expected numbers. We're asking, "Is our data *a good fit* with the theory?" The name of the test actually fits what it does!

Do we have categorical data? We have a count of freshmen, sophomores, etc. Check.

Are we comparing it to real or theoretical data? Our expected "team" of 25% freshmen, 25% sophomores, 25% juniors and 25% seniors is definitely not real. Again, check.

Now, the easiest way to do a hypothesis test is to find the p-value.

p-value and alpha level

There are two new concepts, a p-value and an alpha level. Let's tackle the p-value first.

The p in **p-value** stands for probability. Meaning p can be any value from 0 to 1, including 0 and 1. Therefore, a p-value of .43 is really a 43% probability.

Probability of what? 43% probability of nothing going on, given the data we have. With such a high probability we would fail to reject our initial hypothesis, the null hypothesis, H_0. But how do we decide if the probability is high enough? We compare the probability to an alpha level.

Throughout this course we will set an **alpha level**, or

significance level for the p-value of .05. That's our fence. For this course, any p-value that's above .05 means we fail to reject the null hypothesis, e.g., nothing's going on, the differences we see are real differences, but are due to random variation. Any p-value below .05 means we reject the null hypothesis. **The differences are statistically different.**

Since no lives are on the line in sports statistics, being 5% (.05) certain nothing out of the ordinary is happening — or 95% certain something is — will be good enough. If you have a job in sports statistics there may be times when you want more certainty or are okay with less certainty. But in this class .05 will work since our purpose is to understand how to work with sports statistics.

PRACTICE PROBLEMS

 Let's use another team to ask the question, How well does the class make-up of the team fit what we expect, e.g., that each class makes up 25% of the team? Here's the class make up of the 2010 Springfield College women's basketball team. (F_O is the Observed frequency.)

	F_O
Freshman	6
Sophomore	4
Junior	5
Senior	0

Wow! No seniors! We have an unequal distribution of players. Is this due to random variation or is this significantly different? Perform a statistical test to answer this question, What's going on with the SC women's basketball team?

 The test needed is a chi-square goodness-of-fit test, and it is detailed below. (F_E is the Expected frequency.)

	F_O	F_E	$F_O - F_E$	$(F_O - F_E)^2$	$(F_O - F_E)^2/F_E$
Freshman	6	3.75	2.25	5.0625	1.350
Sophomore	4	3.75	0.25	0.0625	0.017
Junior	5	3.75	1.25	1.5625	0.417
Senior	0	3.75	-3.75	14.0625	3.750
				$\chi^2 =$	5.533

Adding the end column gets us 5.533. That's our chi-square test statistic (χ^2). Minitab will do this step for you and also give you a *p*-value of .14.

Notes: The expected frequency is 25% (because there are four categories, each category should have ¼ of the total). I did 25% times the total number of players, or .25 × 15 = 3.75. Since each class is ¼ of the total team, notice the expected frequency is the same for each (3.75).

Also the *p*-value is a probability, in this case there is a 14% chance that there is nothing going on with women's basketball given those observed frequencies.

Here is how you will answer this question on the exam:

Hypothesis Test

1. **H_0:** F = S = J = SR
2. **H_A:** The classes are not equal to each other.
3. **Test statistic and its value:** χ^2 = 5.533
4. ***p*-value:** .14
5. **Statistical conclusion:** Since the *p*-value > .05, I fail to reject H_0
6. **Interpretation:** There is nothing going on with women's basketball. The differences we see are due to random variation

---

Investigate the 2016 SC Golf and SC Women's XC teams to determine if there is a good fit between their observed frequencies on their rosters and their expected frequencies. You can do this by answering the question, Is there something going on with the golf and cross-country teams.

First, look up the rosters for both teams and enter the data into Minitab as shown below:

	C1-T	C2	C3
	class	observed frequency SC golf	observed frequency womens XC
1	freshman	6	3
2	sophomore	0	6
3	junior	7	4
4	senior	2	6
5			

I choose Stat from the menu bar then Tables, Chi-Square Goodness-of-Fit Test (One Variable) on Minitab, and the results are as follows:

```
Chi-square Goodness-of-Fit Test for Observed Counts
   in Variable: observed frequency SC golf
                                                      Test
        Contribution
Category     Observed   Proportion   Expected    to Chi-
    Sq
freshman         6         .25        3.75       1.35000
sophomore        0         .25        3.75       3.75000
junior           7         .25        3.75       2.81667
senior           2         .25        3.75       0.81667

 N   DF   Chi-Sq   P-Value
15    3  8.73333    .033
```

Then analyze and interpret the data using the following format:

Hypothesis Test

1. **H_0:** F = S = J = Sr
2. **H_A:** The classes are not equal
3. **Test statistic and its value:** Chi-square = 8.73
4. **p-value:** .033
5. **Statistical conclusion:** Since the p-value < .05, I reject H_0
6. **Interpretation:** The differences we see are not due to random variation. There is something going on with men's golf at SC.

```
Chi-square Goodness-of-Fit Test for Observed Counts
   in Variable: observed frequency women's XC
                                                      Test
        Contribution
Category     Observed   Proportion   Expected    to Chi-
    Sq
freshman         3         .25        4.75        .644737
sophomore        6         .25        4.75        .328947
junior           4         .25        4.75        .118421
senior           6         .25        4.75        .328947

 N   DF   Chi-Sq   P-Value
19    3  1.42105    .701
```

Hypothesis Test

1. **H_0:** F = S = J = Sr
2. **H_A:** The classes are not equal
3. **Test statistic and its value:** Chi-square = 1.42
4. ***p*-value:** .701
5. **Statistical conclusion:** Since the *p*-value > .05, I fail to reject H_0
6. **Interpretation:** There is nothing going on with SC women's XC, the differences we see are due to random variation.

TEST YOUR UNDERSTANDING OF CHI-SQUARE GOODNESS-OF-FIT TEST

Perform a chi-square goodness-of-fit test on the 2019-20 Springfield College women's softball roster to determine if the number of athletes, as organized by class, fits the expected model. Answer, Is there something going on with SC softball?

In Minitab, first enter Class in C1 and Observed Count in C2. Categorize our one grad player as a senior. Run your chi-square analysis. Then answer the following 14 questions.

1. What is your null hypothesis (H_0):

2. What is your alternative hypothesis (H_A):

3. Which hypothesis are we testing?

4. Are there differences in the data for softball?

5. What does the null hypothesis attribute those differences to?

6. What-test statistic are we using for the goodness-of-fit test?

7. What is the value of your test statistic?

8. What is the alpha (or significance) level?

9. When you perform the test, can you get a *p*-value > 1?

10. Why or why not?

11. What is your *p*-value for softball?

12. Statistically, what does your *p*-value tell you to do?

13. In the context of the problem, what do your results mean?

14. Is there a good fit between your observed frequencies and your expected frequencies?

Here is my Minitab output after choosing Stat, Tables, Chi-Square Goodness-of-Fit Test (One Variable):

Chi-Square Goodness-... ✓ ✕

WORKSHEET 1

Chi-Square Goodness-of-Fit Test for Observed Counts

Using category names in Class

Observed and Expected Counts

Category	Observed	Test Proportion	Expected	Contribution to Chi-Square
Freshmen	6	0.25	5.5	0.045455
Sophomore	5	0.25	5.5	0.045455
Junior	4	0.25	5.5	0.409091
Senior/Grad	7	0.25	5.5	0.409091

Chi-Square Test

N	DF	Chi-Sq	P-Value
22	3	0.909091	0.823

	C1-T	C2	C3	C4	C5
	Class	Observed count (frequency)			
1	Freshman	6			
2	Sophomore	5			
3	Junior	4			
4	Senior/Grad	7			

1. What is your null hypothesis (H_0): F = S = J = Sr/Gr

2. What is your alternative hypothesis (H_A): F ≠ S ≠ J ≠ Sr/Gr

3. Which hypothesis are we testing? The null hypothesis. H_0. All semester we will only test only H_0.

4. Are there differences in the data for softball? Yes. There are 6 first-years, 5 sophomores, 4 juniors and 7 seniors/grad students.

5. What does the null hypothesis attribute those differences to? Random variation, recruiting, injuries, study abroad, etc.?

6. What-test statistic are we using for the goodness-of-fit test? Chi-square. In Minitab, we choose Stat, Tables, Chi-Square Goodness-of-Fit Test (One Variable).

7. What is the value of your test statistic? .9091

8. What is the alpha (or significance) level? .05 Always .05. All semester we will only use .05.

9. When you perform the test, can you get a p-value > 1? No.

10. Why or why not? The p in p-value stands for probability, and we cannot have a probability greater than 1. It's like a coach asking for 120% effort, we can't do that. If we give 100% effort that is giving everything we have.

11. What is your p-value for softball? .823

12. Statistically, what does your p-value tell you to do? Since the p-value > .05, I fail to reject H_0.

13. In the context of the problem, what do your results mean? There is nothing going on with SC softball, the differences we see are just random variation. Yes.

14. Is there a good fit between your observed frequencies and your expected frequencies?

OPTIONAL: CHI-SQUARE GOODNESS-OF-FIT TEST USING CRITICAL VALUE

The second approach to a hypothesis test is to find what's called the **critical value**. We won't be using critical value in sports statistics class, but it is often used in statistics.

If our test statistic (chi-square in this chapter) is less than this critical value then the variation might be ordinary randomness. If the test statistic is above the critical value then we're fairly confident something is going on to affect our data.

In this course we won't use the critical value. But I'll show you how to find the numbers by hand so you understand what Minitab is doing when it hands us a chi-square test statistic and a p-value.

Q Back to the men's soccer team data. Let's finally get an answer to, "Is something going on with men's soccer at Springfield College?"

Remember I mentioned a test statistic reduces a set of data to a single number? In this chapter our test statistic is chi-square. To turn a set of data into a chi-square value we first ask, "How far is each class from the expected value for that class?"

A The expected value for each class is the 31 total equally spread between the 4 classes, so 31/4 or 7.75. For the freshmen, how far is 7 from 7.75? It's .75. For chi-square it doesn't matter whether the value is smaller or larger. All chi-square cares about is how far the actual value — the **observed value** — is from the **expected value**. (In math terms that's absolute value.)

The second step is to square that value. (That also gets rid of the negatives.)

The third step is to divide by the expected value.

Add all those up and that gives the chi-square test statistic. Here's all the calculations in a table with the chi-square test statistic in the bottom corner. This is relatively easy to do since we only have 4 numbers. It would be an afternoon's work if there were 1000 and we didn't have Minitab.

2012 Springfield College Men's Soccer

Class	Observed freq (obs)	Expected freq (exp)	obs - exp	(obs - exp)^2	(obs - exp)^2/exp
Freshman	7	7.75	-0.75	0.5625	0.07258
Sophomore	10	7.75	2.25	5.0625	0.65323
Junior	9	7.75	1.25	1.5625	0.20161
Senior	5	7.75	-2.75	7.5625	0.97581
Total	31	31	0	14.7500	1.90323

But we have Minitab. In Minitab we'll enter the class names in C1 and the observed frequencies in C2. It will do all the crunching to return the chi-square test statistic. (You'll see that below.)

Okay we have a chi-square value. What do we do with it? In class we won't do anything with it! It's merely a step that Minitab will do for us. But let's peek behind the curtain at what we used to need to do by hand but now Minitab does for us. We will go find a Chi-square Distribution

Table. (There's one in the Appendix.) Here's a part of it relevant to our question.

Chi Squared Distribution Table (partial)

degrees of freedom	Probability of a larger value of χ^2 (*p*-value)									
	0.995	0.99	0.975	0.95	0.9	0.1	**0.05**	0.025	0.01	0.005
1	---	---	0.001	0.004	0.016	2.706	**3.841**	5.024	6.635	7.879
2	0.010	0.020	0.051	0.103	0.211	4.605	**5.991**	7.378	9.210	10.597
3	0.072	0.115	0.216	0.352	0.584	6.251	**7.815**	9.348	11.345	12.838
4	0.207	0.297	0.484	0.711	1.064	7.779	**9.488**	11.143	13.277	14.860

NON-REJECTION REGION	**REJECTION REGION**
Fail to reject the null hypothesis	Reject the null hypothesis

If we cared about different levels of confidence than .05 then all the columns would be important. But remember we said for this course we'll choose a significance level of .05. The bolded column is the only one we care about.

But we're still missing a bit of information. The **degrees of freedom**. That will always be one less than the number of categories we have. Why degrees of freedom? It comes from asking how many values are free to vary. We know the total (31). Once we know how many students are in any three of the classes we know how many are in the fourth class (31 - the first three classes combined). That means three values are free to vary before they lock in the value of the fourth.

Now go to the partial table above. Find the row that corresponds to our degrees of freedom which is 3. Trace over to the (bolded) .05 column. The value there is the critical value, 7.815. If our chi-square value is less than the critical value, it's in the Non-rejection region (colored red). It's labeled "Fail to reject the null hypothesis" which is what we'd do. If it's greater than the critical value (in the Rejection region), we reject our null hypothesis. Our chi-square value is 1.90323. It is less than 7.815. That puts us in the red area. That tells us we fail to reject the null hypothesis. There isn't enough evidence to suggest the variation is anything other than natural.

That's how we use critical value and test statistic in hypothesis testing. Which we won't use again. We will stick with the *p*-value. Let's enter the data into the worksheet.

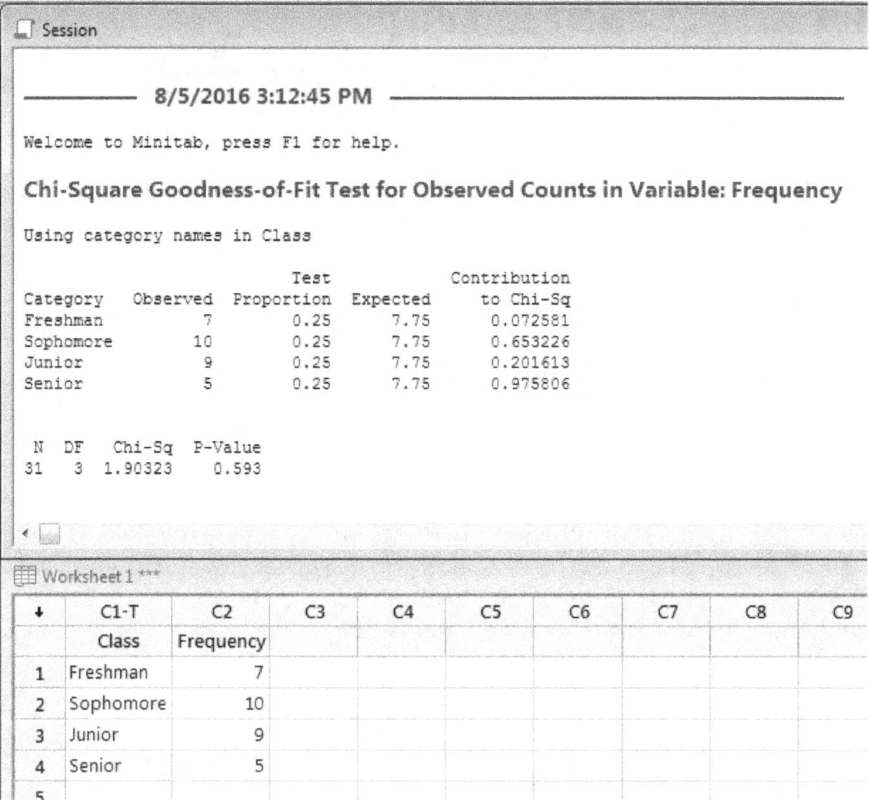

Notice the chi-square test statistic is 1.90323, which is the same result we calculated by hand! Also the *p*-value is .593. Much easier! But now you have an idea of what Minitab goes off to do behind the curtain.

You just completed the ANALYSIS portion of the collect, organize, **analyze** and interpret the data! Congratulations, right? Before we pat ourselves on the back, we have to interpret the results. Remember, interpretation is the most important step.

Step 5 on the hypothesis test will be: "Since the *p*-value > .05, I fail to reject H₀." Step 6 is our interpretation. We would say, "There is nothing going on with the men's soccer team. The differences we see in the data are due to random variation." We compared the 7/10/9/8 numbers we found on the roster to the expected numbers 7.75/7.75/7.75/7.75, which are obviously different. But the differences are out to random variation.

PRACTICE PROBLEMS

Use the 6 steps of the hypothesis test that we learned in Chapter 5 to answer these questions.

Hypothesis Test

1. **H_0:**
2. **H_A:**
3. **Test statistic and its value:**
4. **_p_-value:**
5. **Statistical conclusion:**
6. **Interpretation:**

Perform a chi-square goodness-of-fit test to determine if the number of athletes on the most recent Springfield College women's lacrosse team, as organized by class, fit your expected model. To do that, answer the question, Is there something going on with women's lacrosse?

Women's Lacrosse Roster

Class	Observed Frequency	Expected Frequency
Fr		
So		
Jr		
Sr		

Do the same for the most recent Springfield College men's lacrosse team.

Men's Lacrosse Roster

Class	Observed Frequency	Expected Frequency
Fr		
So		
Jr		
Sr		

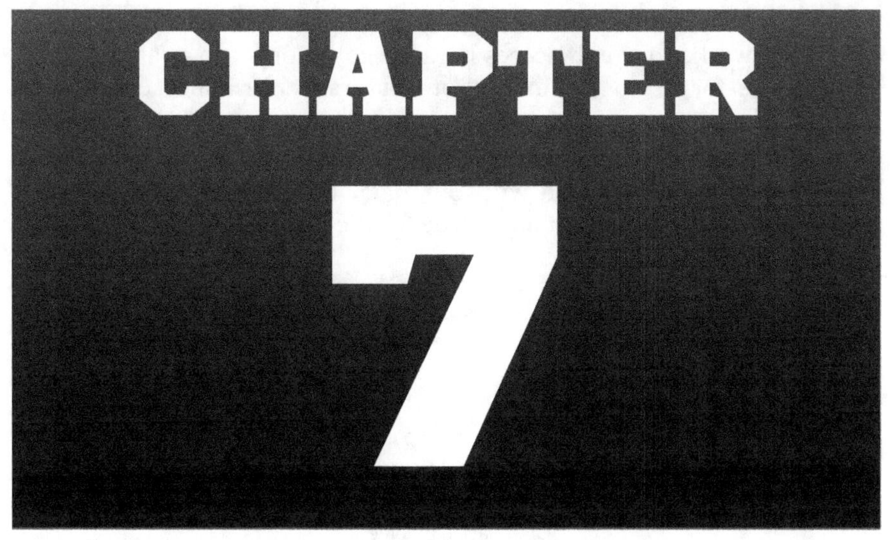

CHAPTER 7

Does average make sense?

Statistics always remind me of a fellow who drowned in a river where the average depth was only three feet.

— Woody Hayes, OSU football coach

PREVIEW: CATEGORICAL OR QUANTITATIVE? USE AVERAGE TO DECIDE

We know there are two types of variables, categorical and quantitative. Categorical variables represented a group or a category. We covered categorical variables in Chapters 1 – 6.

Now in Chapter 7, we address quantitative variables. The rest of the book is quantitative variables. Remember, the root of **quantitative** is quantity. Quantitative variables represent quantities or numbers. They are numerical.

We did quantitative variables way back in day two when we did mini-putt. I took your golf scores from the front nine and the back nine and listed them on the board. Then we used Minitab to calculate the class average for the front nine and the average for the back nine. We termed that average the mean.

It turns out the mean will be used right up until the last day of class! The mean worked for golf scores because golf scores are quantitative. It made sense for us to summarize our class's performance on the golf course with the mean. The mean was a nice, easy to understand measure to describe our class's golf scores.

If a mean makes sense with our data, then the variable must be quantitative. Let's see if that holds true for the 2019 Red Sox offense:

Rank	Pos	Name	Age
1	C	Christian Vazquez	28
2	1B	Mitch Moreland*	33
3	2B	Brock Holt*	31
4	SS	Xander Bogaerts	26
5	3B	Rafael Devers*	22
6	LF	Andrew Benintendi*	24
7	CF	Jackie Bradley Jr.*	29
8	RF	Mookie Betts	26
9	DH	J.D. Martinez	31
10	IF	Michael Chavis	23
11	C	Sandy Leon#	30
12	2B	Eduardo Nunez	32
13	UT	Sam Travis	25
14	2B	Marco Hernandez*	26
15	1B	Steve Pearce	36
16	OF	Gorkys Hernandez	31
17	MI	Chris Owings	27
18	UT	Blake Swihart#	27
19	MI	Tzu-Wei Lin*	25
20	2B	Dustin Pedroia	35
21	C	Juan Centeno*	29

This list came from baseball-reference.com. I copied just the offense for 2019. Baseball Reference ordered them in the first column, 1-21. All columns are variables.

We know there are only two types of variables, categorical and quantitative. Is the first column categorical or quantitative? They're numbers. We can find the mean, right? But does that mean make sense?

To get the mean, I add up the rankings in the first column and then divide the total by 21. That leaves me with a mean of 11 for the first column. The mean of 11 does not make sense. Mathematically, we would never take an average of the order. Why? The mean does not make sense. Even adding the column felt weird. We would never even think of adding rankings. Making the first column is categorical.

The second column is easier, no numbers! Positions are groups or categories, so Position is also a categorical variable. Even the third column, Name, is categorical. Usually, names are categories of one, making them statistically boring. But still a categorical variable.

Finally, in the fourth column, we have Age. Age is a number. Let's find the average age for the 2019 Red Sox offense. We add up the ages and divide the total by 21. We get a mean of 28.38 years.

Does it make sense to say the average *2019* Red Sox offensive player is 28.35 years old? YES! Average, or mean, makes sense. Age is a quantitative variable.

In the next chapter (Chapter 8), we will begin to learn in detail about the graphs for quantitative variables. But here is a sneak peek for you.

Back in Chapter 4, we learned about bar charts and pie charts to picture categorical variables.

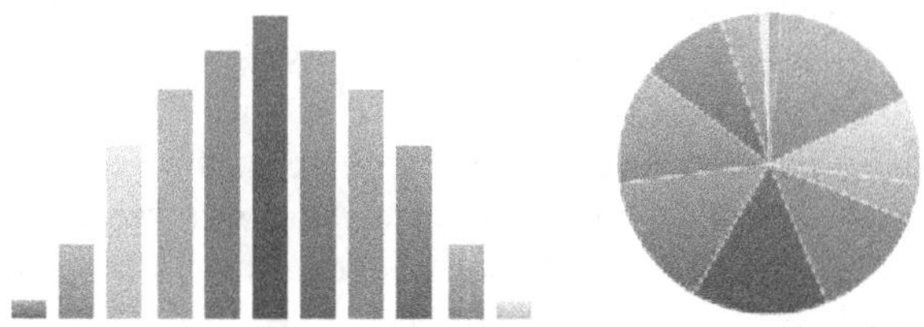

Can't we use them again? No, unfortunately not. Our data now is numerical (continuous). It is no longer discreet as categories are. We will find continuous numbers can be seen more clearly with either a histogram or boxplot. Our two new graphs will be histograms or boxplots. Histograms or boxplots.

Coming up, we will learn to graph a histogram for any problem with numerical data. This is because histograms are easy to visualize. Histograms have a definite look to them.

If a **histogram** (a graph like below) has a nice bell shape, that's the graph to use. Keep the histogram.

Pick Me!

If the histogram looks like anything else …

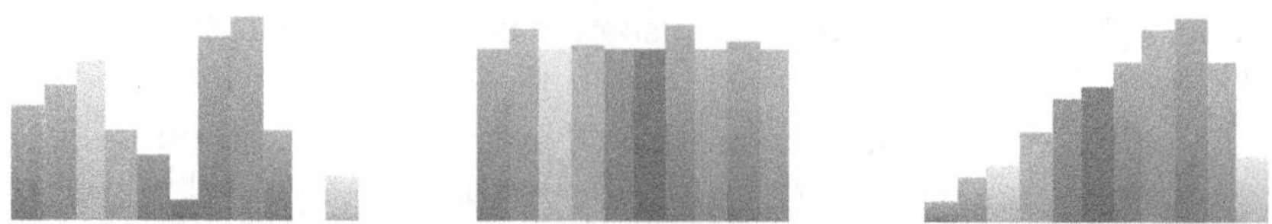

we will switch to a **boxplot**. We will learn about boxplots in detail in Chapter 10, but typically a boxplot looks like this:

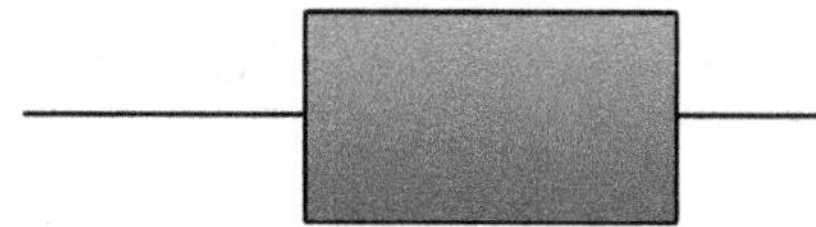

Don't bar graphs and histograms look a lot alike? Yes, they do! Essentially they're the same but are used differently. We use bar charts to compare categorical variables. We use histograms to show variables when they are continuous numbers. The difference — besides gaps between bars on bar charts and none on histogram — is each bar in a bar chart refers to a category. In histograms, the bars refer to numbers. But not just numbers. Groups of numbers. A continuous range of numbers.

Each bar represents a range of values. In the histogram below, the first bar, the 0-4, represents all the Red Sox offensive players in 2019 who hit less than five home runs. There were ten of them. All the way out to the far right which shows just one Red Sox player who hit between 35 and 39 home runs. That happened to be JD Martinez with 36.

Here is the histogram for the 2019 Red Sox hitters' total home runs:

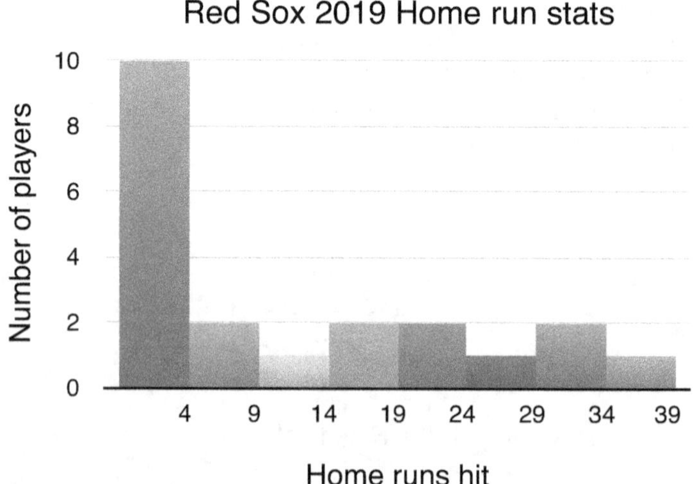

Note: The range won't always be integers. Batting averages might have one bar of .200 to .210 and the next .211 to .220.

In a histogram, the bars refer to numbers. But not just numbers. Groups of numbers. A continuous range. Even though the bars might be labeled with a single number, they represent a range, for example 0-4, 5-9 and so on. The range won't always be integers. It might be batting averages with the range of one bar .200 to .210 and the next .211 to .220. It only makes sense if the bars go in numeric order.

A picture is worth a 1000 words (take two)

They both (statistics & bikinis) show a lot, but not everything.

— *Toby Harrah, baseball infielder*

PREVIEW: EXPLORING MORE GRAPHS TO SUMMARIZE DATA

I mentioned that when quantitative data has a bell shape, we should use a histogram to picture the data. If the data is not symmetric, use a boxplot. It is easy for me to say, but real-life data will rarely make it that easy for us.

To give you an idea of what symmetrical versus non-symmetrical looks like for real data, I entered the batting averages for all of the qualified hitters in 2019 for the National League (NL), American League (AL) and the Major Leagues combined (MLB) into Minitab. Then I had Minitab create three histograms, one for each of the 2019 hitters.

Symmetry means we have most of the data in the middle. Then the data flows down evenly on both sides. Symmetry has a bell shape with no outliers. Non-symmetric is everything else (skewed left, skewed right, uniform, bimodal, trimodal, outlier, etc.).

Here is the histogram for the 2019 AL hitters:

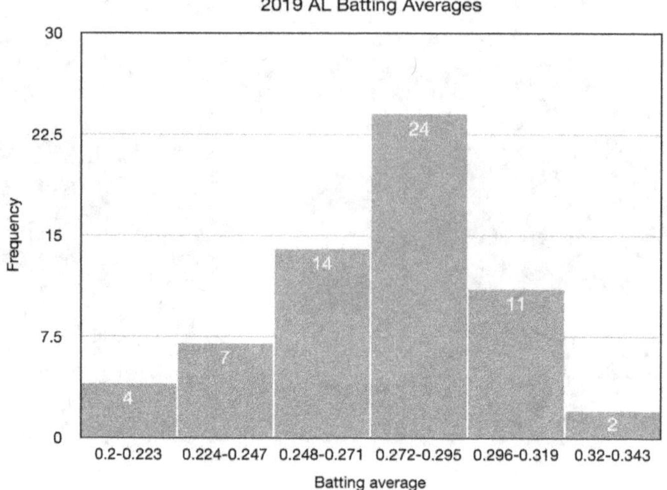

How to read a histogram: the first "bin" goes from .2 - .223, and has a "4" in it. That means there were four hitters in the AL in 2019 who hit between .200 and .223, including our worst hitter, Rougned Odor of Texas, who hit .205.

Read across and the other end of the histogram tells us there were two hitters in the AL in 2019 who hit between .320 and .343. They were DJ LeMahieu of the Yankees (.327) and our 2019 batting champion, Tim Anderson of the Chicago White Sox.

For the Red Sox, Rafael Devers was our best hitter (.311) and he is one of the 11 in the bin between .296 and .319.

How should we describe the AL histogram? It skews to the left. Look at the largest bin (.272 - .295), and notice there are three bins and 25 totals players to the left (which is more than the right), that makes our graph skewed left. It is skewed in the direction of the most data is how we decide left or right.

How about the 2019 NL hitters:

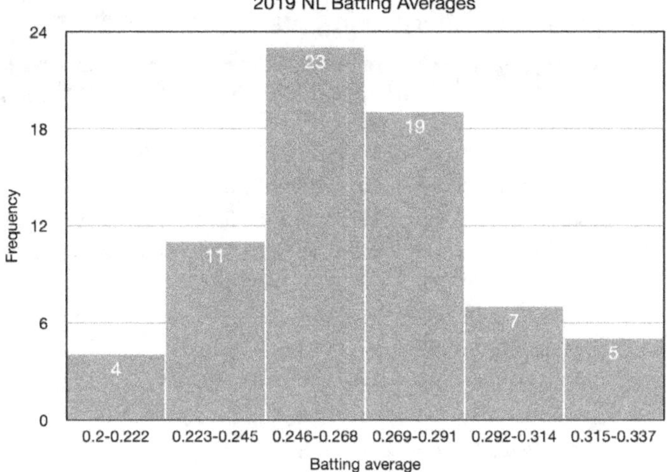

The NL is skewed right. I first looked for the largest bin (.246 - .268) and counted from there. There are two bins and 15 players to the left, and three bins and 31 total players to the right. So skewed right!

And MLB combined in 2019:

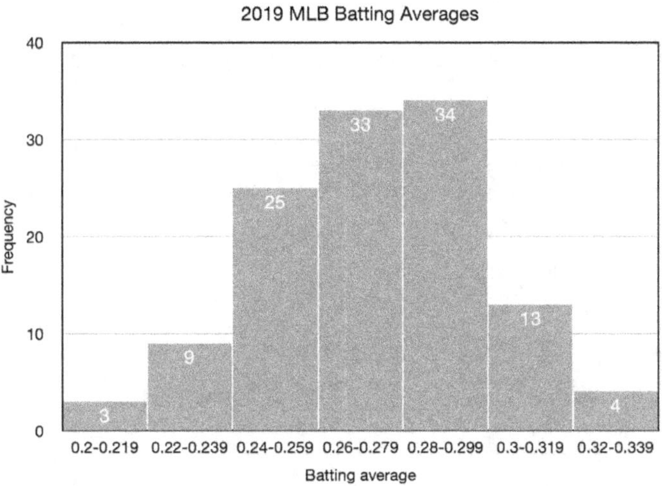

Skewed left.

Let's try one more example. In 2019, Springfield College lead all of Division III in rushing! We were number one in the country. Below are the rushing statistics for the 2019-20 Springfield College football team.

2019-20 Football Rushing Statistics - Springfield

No	Name	Yr	Pos	GP	Rush	Yds	Y/G	Avg	TD	LG	Fum	Lost
11	Chad Shade	Sr	QB	10	182	771	77.1	4.2	8	45	8	4
49	Tim Callahan	Jr	FB	10	121	683	68.3	5.6	6	67	3	0
30	Patrick Ladas	So	FB	10	117	532	53.2	4.5	3	59	-	-
26	Isaiah Cashwell-Doe	So	HB	10	52	399	39.9	7.7	2	60	1	1
28	Ryan Deguire	Jr	HB	10	46	341	34.1	7.4	2	32	1	0
34	Nick Rajotte	Jr	HB	7	33	321	45.9	9.7	4	61	1	1
2	David Wells	Jr	QB	5	27	136	27.2	5.0	2	69	5	2
37	Skyler Bateman	Sr	FB	9	31	111	12.3	3.6	2	10	2	2
24	Jeff Stern	Sr	HB	9	10	95	10.6	9.5	1	25	-	-
36	Jordan Lewis	So		6	8	75	12.5	9.4	0	17	-	-
22	Hunter Belzo	Sr	HB	2	9	58	29.0	6.4	0	16	1	0
19	Shane Brancaleone	So	HB	7	2	19	2.7	9.5	0	18	-	-
38	Nicholas Raneri	Fr	FB	2	3	15	7.5	5.0	0	6	-	-
23	Jamari Jenkins	Fr	HB	1	1	8	8.0	8.0	0	8	-	-
25	Jimmy Shriner	Fr	FB	1	1	5	5.0	5.0	0	5	-	-
2x	Areh Boni	Fr		3	1	-1	-0.3	-1.0	0	0	-	-
	Totals			10	655	3574	57.4	5.5	30	69	22	10
	Opponent			10	295	969	96.9	3.3	14	50	17	10

I created a histogram of column seven, yards gained rushing.

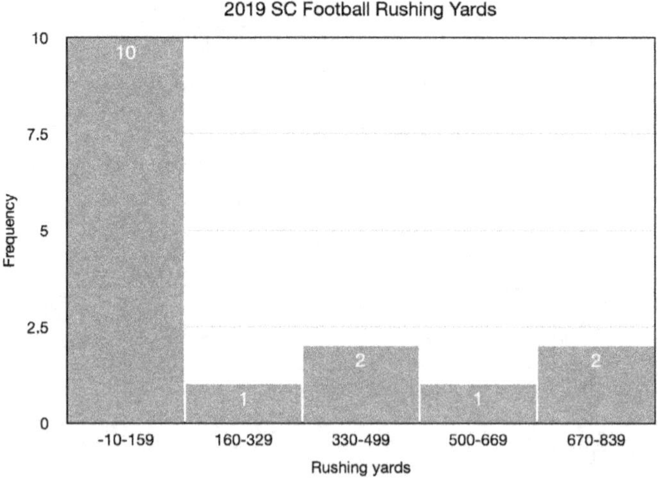

When we analyze our Springfield rushing yards, we can see we are skewed right!

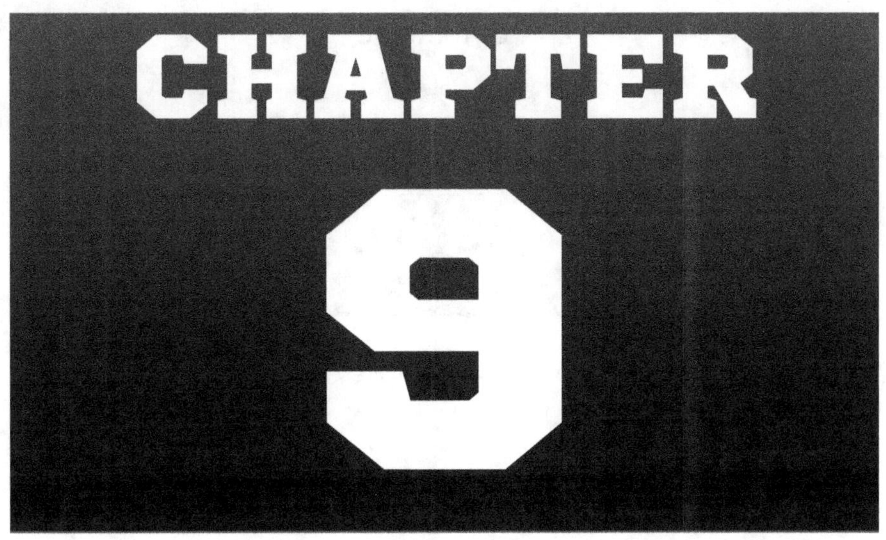

CHAPTER 9

sHAPE, ceNTer & spread

It's like if you and me went for a picnic....We take a chicken and I eat all of it and you have none, but statistically you have eaten half of it....This is why I don't look at statistics.

— *Carlos Carvalhal, Braga manager*

PREVIEW: 3 BASIC QUESTIONS TO UNDERSTAND THE GRAPHS

Before we move onto boxplots, we will investigate how to interpret our data. Once you have created your histogram, you will want to interpret it. When we describe a histogram, we look at the **shape**, **center**, and **spread**.

Shape

We will do shape first. There are two different types of shapes our data can take, symmetric and non-symmetric. The shape is easiest to see when we graph the data as a histogram. In class and then on the exams, graph the data as a histogram first.

Symmetric has a single peak in the middle and is relatively bell-shaped with no outliers.

Symmetric

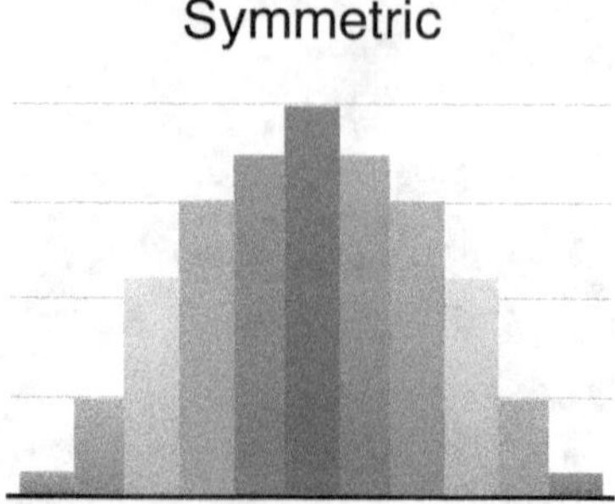

Non-symmetric is ... any other shape! Here are two examples of non-symmetric distributions.

Bimodal
(two peaks)

Outlier

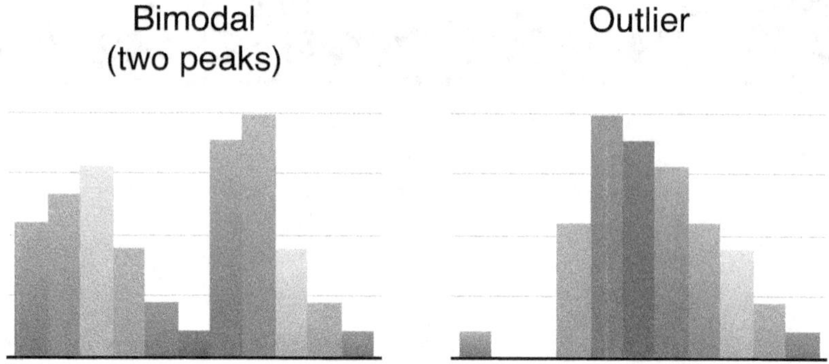

There are even more non-symmetric than bimodal and outlier. The data can be trimodal or multimodal. It can be random. Have gaps. Have outliers. Anything that isn't symmetric or fits under a bell-shaped curve is classified as non-symmetric.

Skewed (pulled) left
(tail to left)

Skewed (pulled) right
(tail to right)

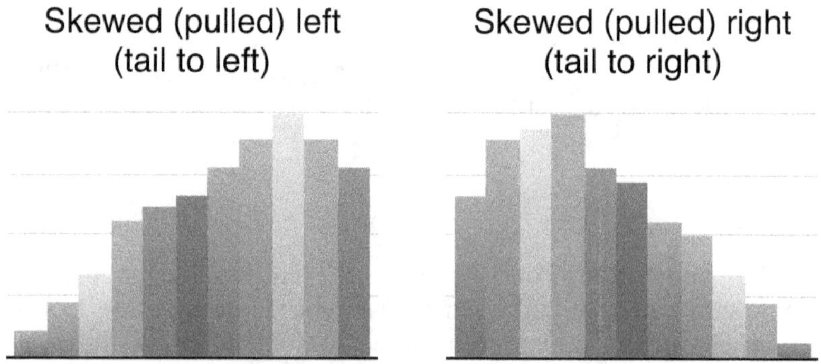

In terms of the shape, center, and spread, that is all we need to know about shape.

Center

What about the center? Well, that is almost as easy. The center is in the center of the graph. It doesn't matter whether the graph is a histogram or a boxplot. The center is the center.

But we use different descriptive stats to describe the center. For **symmetric** data, represented by a histogram, we use the average of the data. In stats, averaging the data is called the **mean**.

Non-symmetric data affects the mean, pulling it either left or right, depending on the data. We cannot use the mean for non-symmetric data. But a measure of the center not affected by non-symmetric data is the **median**. The median is the middle value when you put the data in order from low to high.

> **Symmetric goes with mean.**
>
> **Non-symmetric goes with median.**

Minitab will also give us the mode, a third measure of the center. The mode is just the number that occurs most often. I will show you the mode below, but it has limited value in sports examples.

If we have a graph that has a relatively symmetric distribution with one peak in the middle forming a bell-shaped curve, we use the mean.

As we defined it earlier, the mean is just the statistics word for the average. To calculate the mean by hand, you add up all the numbers and divide by how many numbers you have. But let the Minitab do the drudgery for you. We want to concentrate on interpreting the result without getting bogged down with calculations. You all have access to Minitab. Let it do the grunt work. You can focus on interpreting those numbers.

If our graph is non-symmetric — that is looks like anything other than a bell shape — we use the median to show the center of the distribution. The median is the middle number in a distribution of numbers.

For instance, Tom Brady had quarterback ratings of 88.6, 111.0, 88.4 in the last three weeks of the 2019 season. His average, or mean, was 96, which I got by adding up the three ratings and dividing by three. His median rating was 88.6, the middle number in terms of putting them in order of lowest to highest, then picking out the middle number.

> **Original order:** 88.6, 111.0, 88.4
>
> **Low to high order:** 88.4, **88.6**, 111.0

Which is why Tom's median is 88.6 yards.

Both numbers — the mean and the median — described the center, but remember, we use the mean with a symmetric distribution and the median with everything else. Center is the typical value. For Tom, there is a big difference between saying he ended his career in New England with a typical QB rating of 96.0, or with 88.6. Unfortunately for Tom, his QB rating is non-symmetrical, we have to use the lower median rating of 88.6 to summarize his last three weeks.

There's one exception. If you have a symmetric distribution, but there is an outlier, The outlier pulls the mean towards it. The mean is overly excited by the outlier and gravitates towards it. In those cases the median is the best way to measure the center of the data. The median can resist the outlier's charms and stay put. If you have outliers, use the median.

Spread

That takes care of shape and center. Now we tackle the spread of the data. The spread gives us an indication of how wide apart the data is. If the data is bell-shaped — **symmetrical** with no outliers — use the **standard deviation** to measure the spread.

I will mention standard deviation now, but explain the concept in the next chapter. For now, click Minitab and everything will show up on one screen:

Descriptive Statistics: ...

WORKSHEET 2

Descriptive Statistics: 2019 Pats rushing yards

Statistics

Variable	Mean	StDev	Median	IQR
2019 Pats rushing yards	121.6	247.2	24.0	112.3

If you have data that is non-symmetrical, use the interquartile range (IQR) for the spread. What about the basic range, you ask? The problem with range is it is based only on the highest and lowest values. Those values might be very large or small that they don't represent your distribution. We use the IQR, which is Q3 – Q1.

What are these Qs? Well, it's like a basketball game broken down into quarters. Q1 is the first quartile (the end of the first quarter), Q3 is the third quartile (end of the third quarter). Q3 - Q1 gives us the middle 50% (75% - 25%) of our data.

Most of the time, that middle 50% is much more indicative of the distribution than the range is. The middle 50% of the data eliminates any extremes from our analysis. So, the IQR = the interquartile range = Q3 − Q1 = the middle 50% of the data. That is the value we use most often to show the spread of the data when it is not symmetrical.

GUIDED EXAMPLE FOR SHAPE, CENTER AND SPREAD WITH THE PATS

Find the shape, center and spread of the 2019 rushing yards for the New England Patriots.

I looked up and copied the 2019 rushing yards for the Patriots here:

2019 Patriots stats

Name	Pos	Rushing yards
Sony Michel	RB	912
Rex Burkhead	RB	302
James White	RB	263
Brandon Bolden	RB	68
N'Keal Harry	WR	49
Tom Brady	QB	34
Julian Edelman	WR	27
Phillip Dorsett II	WR	21
Damien Harris	RB	12
Mohamed Sanu Sr.	WR	8
Antonio Brown	WR	5
James Develin	FB	3
Josh Gordon	WR	1
Jarrett Stidham	QB	-2

Then I loaded rushing yards into C1 then graphed a histogram.

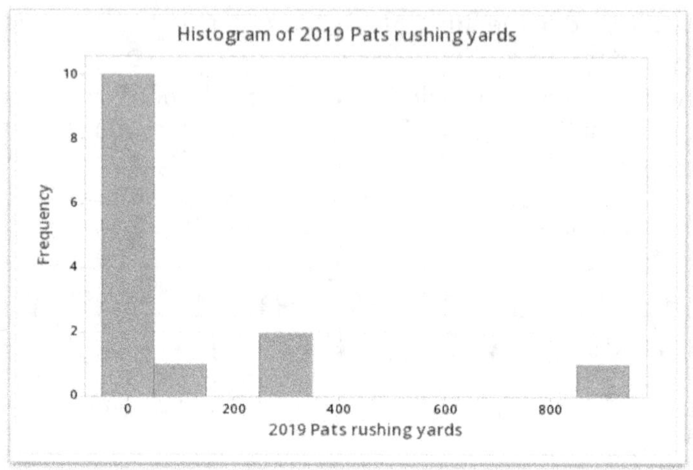

I use this histogram to determine my **shape** which is skewed right with an outlier, therefore **non-symmetric**.

I had Minitab calculate the center and spread for symmetric and non-symmetric for teaching purposes. Here is that Minitab output:

Descriptive Statistics: ...

WORKSHEET 2

Descriptive Statistics: 2019 Pats rushing yards

Statistics

Variable	Mean	StDev	Median	IQR
2019 Pats rushing yards	121.6	247.2	24.0	112.3

Since the graph is non-symmetric I use the median and IQR to describe the center and the spread. The center is 24 yards. The spread is 112.3 yards.

Shape: non-symmetric

Center: 24 and,

Spread: 112.3

While we're here, Sony Michel wanted me to notice how his rushing yards were three time better than his next-best teammate Rex Burkhead. His 912 yards is far, far to the right. That makes Michel's value an **outlier**.

When data is symmetric, the mean and median are the same. Without Michel's yards, the Patriots the mean and median would both be 24 yards. His 912 yards PULLED the mean up to 112.3. Outliers pull the mean. They can make the group's average look much better — or worse! — than it actually is.

On the other hand, notice the median stayed at 24, even with Michel's 912 yards in the calculation. The median is resistant to outliers. That is why we use the median with outliers and non-symmetric data in general. 24 yards is a better measure of the center of the data since it is a more typical number for the 2019 Patriots rushing offense.

PRACTICE PROBLEM

Below you will find some data. First, I **COLLECTED** my son Danny's 5K times. Then I **ORGANIZED** it into a file from the earliest race to his latest. Now we can **ANALYZE** the data. Then we will **INTERPRET** the data.

First, let's look at the data:

Danny's 5K Times

Race	Time
Super 5K	23:56
Day of Portugal	24:07
Rhody 5K	25:32
Mass Senior Games	25:38
Run for Humanity	25:44
Norfolk Jingle Bell Run	25:47
Norfolk Dare	26:40
Angino Race	26:46
Gilio Memorial Race	27:22
Foxborough Flat 5K	27:30
Great Bear Run	27:30
Medway Lions Road Race	27:47
Team Hoyt	27:58
Bone Density Dash	28:23
Icicle Series	28:30
Christopher's Run	29:12
Attleboro YMCA	29:15
Icicle Series	29:15
RI Jingle Bell Run	29:37
Icicle Loop	29:54

Race	Time
Walter's Run	29:58
Icicle Series	29:59
Holyoke Elks	30:07
Icicle Series	30:30
Icicle Loop	30:37
Icicle Series	30:46
Icicle Loop	31:26
Icicle Series	32:51
Icicle Series	33:44
Icicle Series	34:50

Then answer these questions.

1. Create a histogram and a boxplot for Danny's 5K times.

2. What is the shape of the distribution of his race times?

3. What is the mean? Median? Standard deviation? IQR (Interquartile Range)?

4. Which measure is appropriate for the center of the distribution? Why?

5. Which measure is appropriate for the spread of the distribution? Why?

6. INTERPRET the data.

Answer to practice problem

Entering the data above into Minitab, here are answers to each step of the question.

1.   Create a histogram and a boxplot for Danny's 5K times.

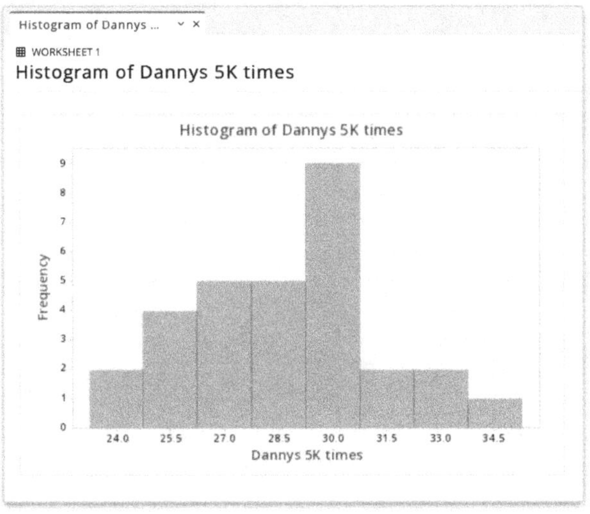

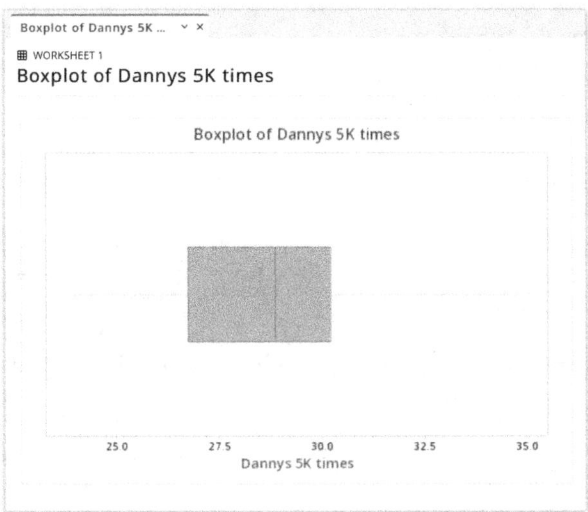

2.   What is the shape of the distribution of his race times?

For the shape, it is easier to see from the histogram that the data skews left. There are more bins and individual races to the left.

3. **What is the mean? Median? Standard deviation? IQR (Interquartile Range)?**

Using the descriptive statistics from Minitab we can answer:

Descriptive Statistics: ...   ⌄ ✕

▦ WORKSHEET 1

Descriptive Statistics: Dannys 5K times

Statistics

Variable	N	N*	Mean	StDev	Minimum	Q1	Median	Q3	Maximum	IQR
Dannys 5K times	30	0	28.706	2.638	23.933	26.742	28.850	30.213	34.833	3.471

Variable	Mode	N for Mode
Dannys 5K times	27.5, 29.25	2

	C1	C2	C3	C4	C5	C6	C7	C8	C9
	Dannys 5...								
1	23.9330								
2	24.1167								
3	25.5330								
4	25.6333								

- What is the mean? 28.706
- Median? 28.850
- Standard deviation? 2.638
- IQR (Interquartile Range)? 3.471

4. **Which measure is appropriate for the center of the distribution? Why?**

Because the data is skewed, the more appropriate measure is the median of 28.850 minutes

5. **Which measure is appropriate for the spread of the distribution? Why?**

Because the data is skewed, we use the IQR of 3.471 minutes

6. **INTERPRET the data.**

Danny's typical 5K time was 28.850 (28:51) minutes with a spread of 3.471 (3:28) minutes.

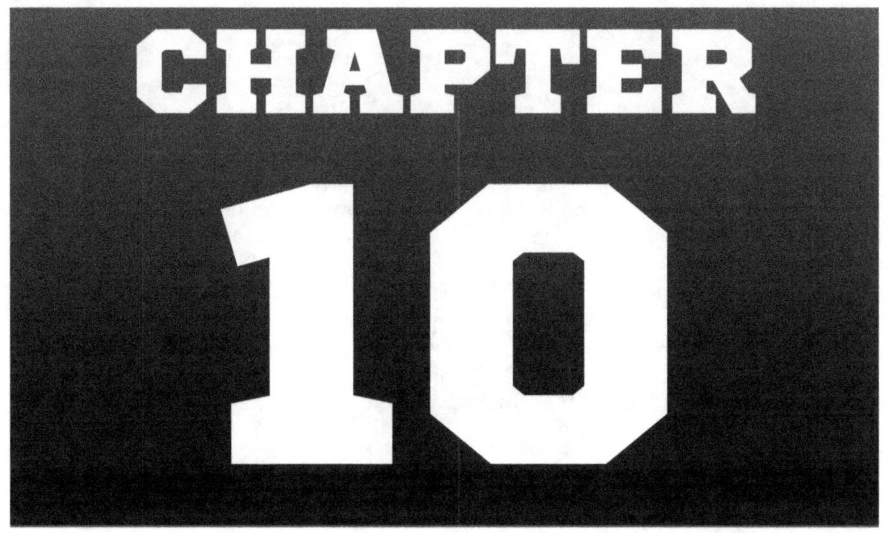

A box with whiskers?

Baseball is 90 percent mental. The other half is physical.

— *Yogi Berra, American icon*

PREVIEW: A NICE EASY GRAPH BASED ON THE 5-NUMBER SUMMARY

When you have quantitative data, look at the histogram first. You will quickly see if the data is symmetrical or not. If it is symmetrical (or relatively so), then keep using the histogram! If not, for any of the reasons we explored in the last chapter, then switch to a boxplot.

A boxplot draws a box around the middle half of the data, then extends lines (whiskers) out left and right to the endpoints of the data. If a data point is an outlier (an extreme value separated from the rest), the boxplot will identify it for you with an asterisk. A boxplot can have multiple outliers (asterisks) or one or none.

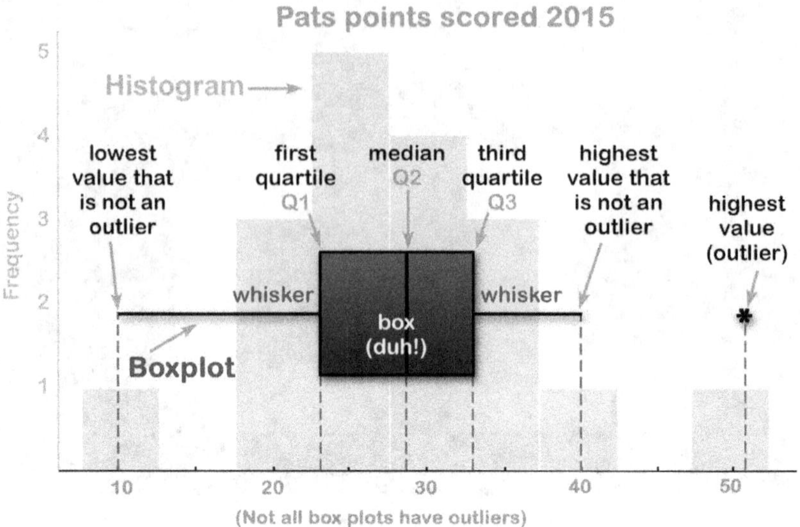

GUIDED EXAMPLE FOR BOXPLOTS WITH THE AMERICAN LEAGUE

Enter data in Minitab

I took the batting averages for qualified hitters in the American League in 2015. I loaded them into Minitab in C1.

Choose Minitab options

I chose Graph from the menu bar and then Boxplot. With box plots I want to click Scale, Transpose, then click OK.

Done!

Here is the boxplot:

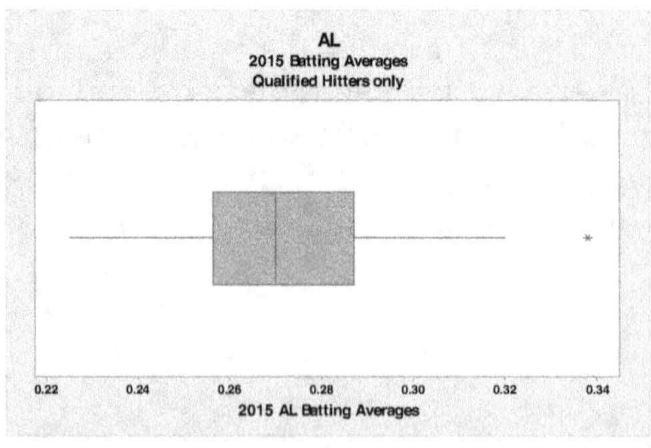

Notice in the graph, the boxplot has a high asterisk, which turns out to be a dot for Miguel Cabrera. This asterisk highlights to us that he is an outlier. He is the only one. The second-highest average in 2015 was Xander Bogaerts of the Boston Red Sox, who hit .320. But his average was close enough to the rest of the AL that he was not an outlier, so the whisker goes out to .320 to include Bogaerts, and then stops.

The boxplot has five numbers needed to construct it.

- the **lowest** value (minimum)
- **Q1** (first quartile)
- **median** (second quartile)
- **Q3** (third quartile)
- the **highest** value (maximum)

These five numbers are known as the **five-number summary**.

The **minimum** is the lowest value among the 2015 AL batting averages. In our case, for the AL, Logan Morrison of Seattle, and his (not) whopping .225 average.

Q1 is short for the **first quartile**, or the batting average that cuts off the bottom 25% of the data. In 2015, it was Kansas City Royals shortstop Alcides Escobar and his .257 average.

The **median** is the middle number in the data or the 50th percentile. Who was right in the middle of the 2015 AL data? Evan Longoria with his .270 average for Tampa Bay.

Q3 is the **third quartile** or the batter who hit better than 75% of his fellow players. In 2015 that was Adrian Beltre of the Texas Rangers and his .287 batting average.

The **maximum** is the highest average, which brings us back to Miguel Cabrera and his .338.

Plot the five-number summary (Min, Q1, Median, Q3, and the Max) on a number line, put a box around Q1 to Q3, draw a line through the median, and then whiskers out to the minimum and maximum. Calculate to see if any of your data points are outliers, and you have a boxplot! Well, what about that calculation for outliers?

Outliers

What makes Miguel Cabrera's average an outlier rather than part of the data like Xander?

There isn't a single definition of what an outlier is. It just means his average falls outside a "comfort zone." Statisticians want some way to be alerted to odd-looking data. It's data that may be wrong or so unusual that it will skew their analysis of the rest of the data. In fact, there are several ways to identify data that may be odd. We'll just look at the easiest and quickest known as Tukey's range test.

Tukey's test is very simple.

1. We measure the distance from Q1 to Q3. We get that by subtracting the lower value, Q1, from the higher value, Q3. Q3 − Q1 has a special name. It is known as an **Interquartile Range (IQR)** Get it? Inter quartile. It is the interval between the quartiles, between Q1 and Q3. We will talk more about the IQR later on.

2. Then we multiply the IQR by 1.5. This is called the 1.5IQR rule.

1.5IQR rule = 1.5 × (Q3-Q1)

3. To get the "comfort zone" we add that answer to Q3 to construct a high fence. Then we subtract that same value from Q1 to construct a low fence.

4. Any values above the high fence and below the low fence are outliers.

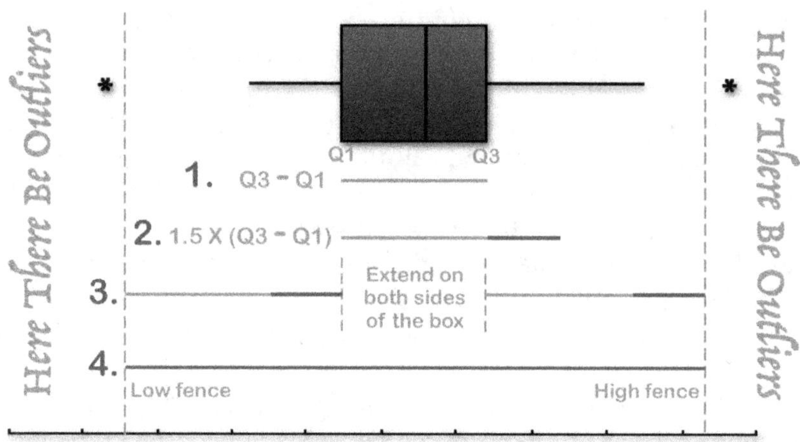

For the AL in 2015, I calculated the IQR at .030 (.287 - .257). Then I multiplied the IQR by 1.5 (.030) and added that to Q3 to get .332. That is my fence. Batting averages below that fence are part of the data. The entire AL was below, except Miggy. His .338 was above that fence, making him an outlier. Below is the Minitab output again for the boxplot for the AL. Notice Minitab automatically calculates the outlier for us and labels it with an asterisk. Thanks, Minitab!

GUIDED EXAMPLE FOR BOXPLOTS WITH SC FOOTBALL

Let's go back to the Springfield College rushing yards data that we looked at in Chapter 8.

Enter data in Minitab

Enter your data in C1. (See below.) (Note, the order of the choices is slightly different on the PC and online versions of Minitab.)

Choose Minitab options

Then go to Graph and pick Histogram... from the drop-down menu. Choose Simple under One Y Variable. Double click on the data value on the left that you want to graph (C1 in this case). For the Option, always keep it simple with Fit normal distribution. Click OK. Online, click Labels and enter a title and subtitle to help your readers to understand the context of your graph. Subtitle 2 will be your name.

Done

Here is what it will look like:

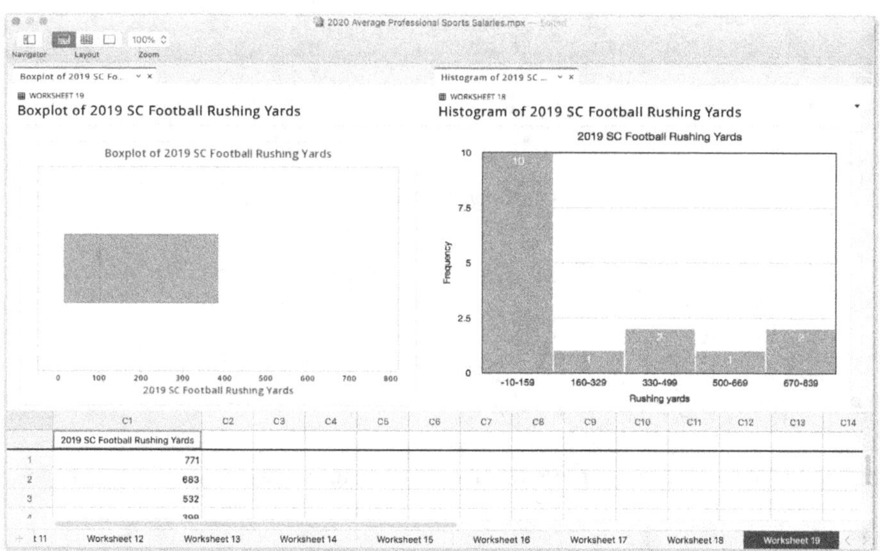

Because the data in the histogram is not symmetrical (remember, it is skewed right in the direction of the tail) we need to do a boxplot. Even though I showed them both in the example above, all you need to do is print out the appropriate graph, in this case, the boxplot.

Notice Josh Carter with his 1549 yards is an outlier, a pretty strong outlier. Not only was the data skewed to the right, it had an outlier too. Either one of those facts alone would have told us to use a boxplot, but both combined confirm we need a boxplot.

PRACTICE PROBLEMS

Mookie Betts was one of the best players ever to play for the Red Sox. And a real nice guy! We will use his career home runs with the Sox to help illustrate boxplots. Mookie's home runs per season:

Year	Mookie's Home runs
2014	5
2015	18
2016	31
2017	24
2018	32
2019	29

Now answer the following questions:

1. Create a boxplot for Mookie Betts' career home runs with the Boston Red Sox.

2. Find the 5-number summary for Mookie:

    - the **lowest** value (minimum)
    - **Q1** (first quartile)
    - **median** (second quartile)
    - **Q3** (third quartile)
    - the **highest** value (maximum)

3. Between what two values do the middle 50% of his home runs fall?

4. Find his IQR.

5. In the data, Mookie hit only five home runs his rookie year. (He only played 52 games.) Is the 5 an outlier?

Create a boxplot for Mookie Betts' career home runs with the Boston Red Sox.

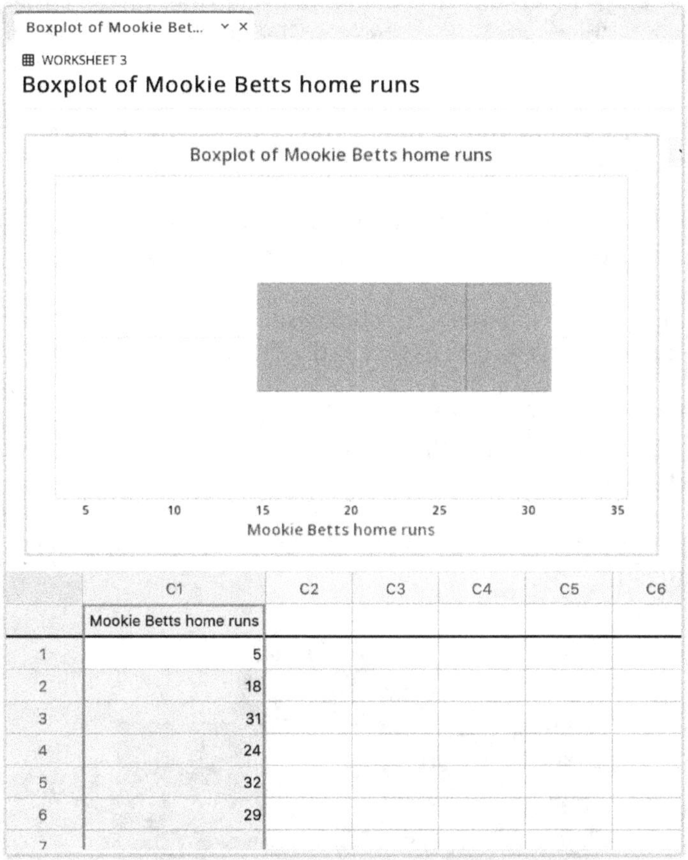

	C1	C2	C3	C4	C5	C6
	Mookie Betts home runs					
1	5					
2	18					
3	31					
4	24					
5	32					
6	29					
7						

6. **Find the 5-number summary for Mookie:**

- the **lowest** value (minimum) = 5
- **Q1** (first quartile) =14.75
- **median** (second quartile) = 26.5
- **Q3** (third quartile) = 31.25
- the **highest** value (maximum) = 32

7. **Between what two values do the middle 50% of his home runs fall?**

14.75 and 31.25, or 15 and 31. This means a typical season for Mookie with the Red Sox was hitting between 15 and 31 home runs.

8. **Find his IQR.**

Remember, IQR stands for the interquartile range. Inter means between, "between." So interquartile is "between the quarters". Between Q1 and Q3.

Mathematically I do Q3 − Q1 to find out what is between them. Or 31.25 − 14.75. Or 16.5!

9. **In the data, Mookie hit only five home runs his rookie year. (He only played 52 games.) Is the 5 an outlier?**

No. I can tell by looking at the boxplot and the far left value (5) is not an asterisk. Therefore, it is part of the data.

Doing it by hand, a low outlier is below Q1 - 1.5IQR. Plugging in our values that means an outlier must be below 14.75 - (1.5 x 16.5) = 14.75 - 24.75 = -10.

Is Mookie's 5 home runs below the fence of -10? No, therefore, it is not an outlier.

CHAPTER 11

Back-to-back boxplots

The triple-double is just a stat. It's a test of your strength and stamina and playing ability, really.

— Oscar Robertson, basketball All Star

PREVIEW: THE GRAPH IS DOUBLY EASY WHEN WE HAVE TWO OF THEM

Boxplots give a good picture of non-symmetrical quantitative data. Back-to-back boxplots allow us to make a quick and accurate comparison of two sets of quantitative data. For instance, the first decade of the 21st century gave us two of the greatest quarterbacks of all time, Peyton Manning and Tom Brady. Who was the better passer in terms of passing yards? Here are the back-to-back boxplots for Tom vs. Peyton:

When we compare boxplots, we only use the box, not the whiskers. Remember,

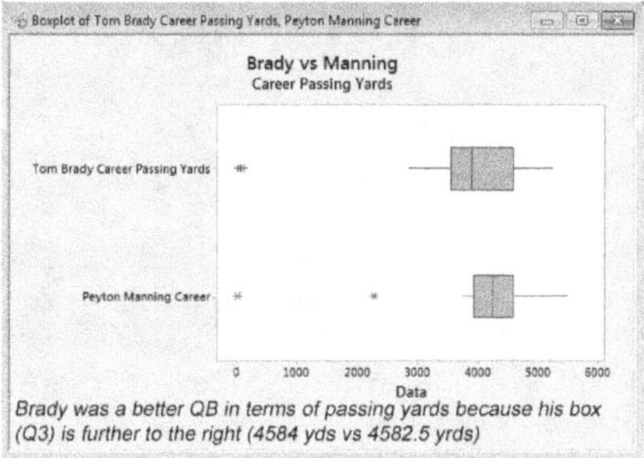

Brady was a better QB in terms of passing yards because his box (Q3) is further to the right (4584 yds vs 4582.5 yrds)

we use a boxplot when the data is not symmetrical. The data outside the middle 50% can be unusual. Like Robel Kiros Habte's 1:04.95 finishing time in the 100-meter freestyle swim in the 2016 Olympic. It was nearly 36% slower than the first-place finisher.

Therefore we are leery of the ends, the first 25% and the last 25%. But we are comfortable with the middle 50%. That is the typical quantity. Since the box goes from Q1 to Q3, just like a basketball game, it represents the middle 50% of the data. If we compared Babe Ruth with Barry Bonds, a boxplot would eliminate the high data points of anyone who used performance-enhancing drugs. And it would eliminate the low end which might be data from injured players.

We just compare the boxes, and look to see which box is further to the right. Specifically, look at the right end of the box. The line for the right end of the box is Q3. Look to see which Q3 is further to the right for Brady and Manning (Brady was). If we were doing Olympic swimmers, it would be which box is further to the left, that is, which Q1 is lower. Brady and Manning want MORE yards, and swimmers want LOWER times.

GUIDED EXAMPLE FOR BACK-TO-BACK BOXPLOTS WITH SC SOCCER

To learn Minitab, we will answer, Which Springfield College soccer team in 2019 was the higher scoring team, the men or the women?

2019 SC Soccer goals

Women	Men
2	1
1	0
1	1
9	3
1	0
1	0
0	2
2	2
6	1
4	0
1	2
7	0
0	1
5	3
3	0
0	2
1	0
1	0
0	
0	
1	
3	
2	
0	

Enter data in Minitab

I entered data for the men in C1 and the women in C2. The data I entered (right) is the number of goals scored for each team.

Choose Minitab options

Click on Graph in the menu bar, then Boxplot. Then highlight the option of Multiple Y's Simple. Enter on OK. Click on C1 Men and they will show up in the box for Graph variables. Then click on C2 Women and they will join them. Then click on Scale and select the box at the bottom for Transpose values and category scales. Then OK. Then Labels and type your Title in the Title box. Then put your name in the Footnote 1 box. Then OK. Then OK again.

Done!

Your output will look like this:

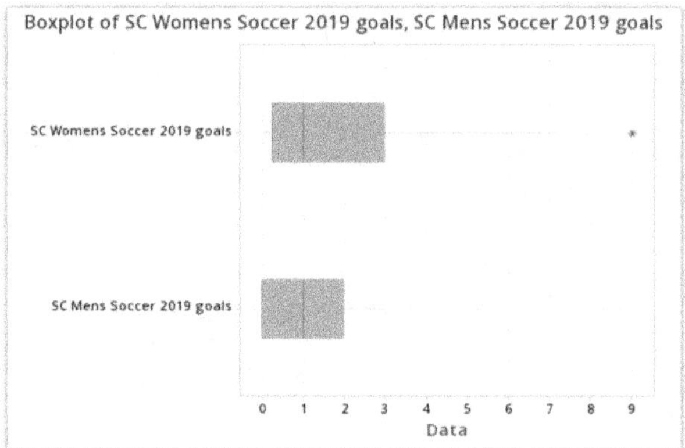

In our example, women's soccer is further to the right because their box is further to the right, making women's soccer the higher scoring team. On the exam you would answer that in the footnote of the graph.

Note for soccer, the women scored a season high nine goals once, and the boxplot identifies that game as an outlier. The men twice hit three goals, so their whisker goes all the way out to three. There is no asterisk because three is not an outlier.

PRACTICE PROBLEM

Albert Pujols and David Ortiz are two of the best home run hitters in baseball. Here are the home runs hit through 2019. Use back-to-back boxplots to answer the question, Which one has had a better career?

Albert Pujols Career HR	David Ortiz Career HR
37	1
34	9
43	0
46	10
41	18
49	20
32	31
37	41
47	47
42	54
37	35
30	23
17	28
28	32
40	29
31	23
23	30
19	35
23	37
	38

Standard deviation in detail

Lajoie was perhaps the first player to attract nationwide attention because of his batting average. Prior to his time, the fans were only mildly interested in statistics.

— Bob French, sportswriter

PREVIEW: MLB HITTERS DEMONSTRATE THE 68/95/99.7 RULE

Standard deviation. We have seen the term before, but may not have totally grasped the concept. Let's try.

To deviate is to depart from a usual measure. When we deviate, we're off track. We've drifted away from where we're expected to be or from some norm.

In statistics the "usual measure" is the mean (the average). The mean is a measure of the center of the data. Each data point is greater than, less than, or sometimes equal to the mean. Each point deviates from the mean. (If the data point is equal to the mean the deviation is 0.)

Here are some examples of deviations. The average height of men in the US is 5 feet 9 inches. I am 6 feet tall. I deviate from the mean by +3 inches. Tom Brady, at 6 feet 4 inches, deviates by +7 inches. Both are a positive deviation since both of our heights are *more than* the mean.

The average height of the 2015 Patriots roster was 6 feet 2 inches. My deviation from the Patriots mean is -2 inches. Tom Brady's deviation is +2 inches. My deviation is negative because I'm *less than* the mean for Patriots' heights.

To do something more interesting, we can ask whether the average deviation of everyone on the team is small — they're all close to the mean — or large — they're very far from the mean. How spread out are they?

That measure is called the **standard deviation**. When we look at all the points — all the team members, all the runs scored, all the football inflations —- we want to find the average deviation from the mean. That average is called the standard, it is the standard deviation from the mean. It is a measure of how far apart the data is. Therefore, standard deviation measures the spread of the data.

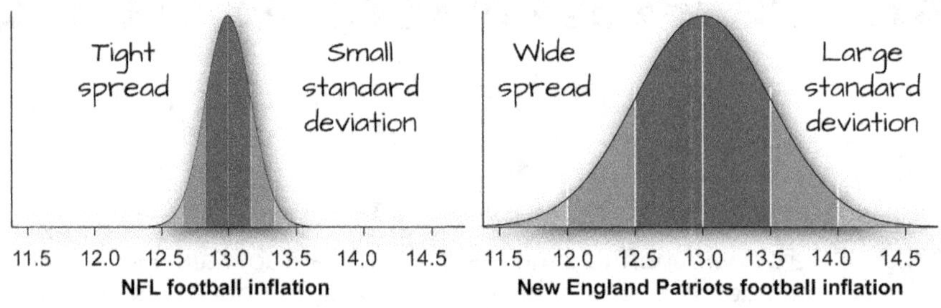

Here's a real example. In the 2010 pre-season, Tom Brady played in all four games and threw for 67, 85, 273, and 51 yards. Let's find his mean and standard deviation.

To find his mean, first we add up all his yards,

67 + 85 + 273 + 51 = 476

Then we divide the total yards by the number of games, 476/4. That's his average yardage per game. We call that his mean. The mean for Tom is 119 yards per game. That is a measure of the center of the data.

$$\textbf{Mean} = \frac{67 + 85 + 273 + 51}{4} = \frac{476}{4} = 119 \text{ yards}$$

We want to see how far each of Tom Brady's games deviated from the mean. Some games he was above the mean, some games below the mean. The second column of the next table came from subtracting his yards for that game minus his mean of 119.

Yards	Yards - Mean
67	-52
85	-34
273	154
51	-68
476	0

Unfortunately, when we add all the numbers in column two, they add up to 0. How can that be? We can see Brady's yards vary from the mean, some more than the mean, some less. The mean is like the balance point. It's the point where the numbers to the left balance the numbers to the right. They will *always* add up to 0.

We need to fix that column so it will not add to zero. To get a number we can do something with, we do a mathematical trick. We square each number to get rid of the negatives. Those numbers are in the third column.

Tom Brady's 2010 Pre-season Yards

	Yards	Yards - Mean	(Yards - Mean)^2
	67	-52	2704
	85	-34	1156
	273	154	23716
	51	-68	4624
Total	476	0	32200
Mean	119		

Now, adding up the third column, we get 32,200. Not zero! We can divide that total by the 4 games to get the average, or 32200/4, which is 8050. That is called our **variance**.

Variance = 32,200/4 = 8050

The variance helps us see patterns in the data. But it isn't a measure of anything physical. In our example, the units of variance would be square yards which has nothing to do with what Tom Brady throws. To fix that, take the square root of the variance. That will get us back to yards. In this case, the square root of 8050 which is 89.72 yards. That is our standard deviation!

Standard deviation = $\sqrt{8050}$ = 89.72 yards

If we had thought about it, we would have expected a large standard deviation because we noticed Brady's yardage varied from 51 to 273, which is a lot. Because of Brady's caliber as a passing quarterback, the preseason games may focus on other skills. We would expect a lot of variation from Brady in the preseason. We would expect his regular-season numbers to be more consistent and therefore have a smaller standard deviation. Let's see.

In the first three games of the 2010 season, he threw for 258, 248, and 252 yards. Remarkably consistent. We would expect a tiny standard deviation as his yardage totals have very little variation. His mean in the regular season in 2010 through three games was 252.7 yards per game.

$$\textbf{Mean} = \frac{258 + 248 + 252}{3} = \frac{758}{3} = 252.7 \text{ yards}$$

Let's calculate the standard deviation using the same process as for the pre-season yards:

Tom Brady's 2010 Yards

	Yards	Yards - Mean	(Yards - Mean)^2
	258	5.33	28.44
	248	-4.67	21.78
	252	-0.67	0.44
Total	758	0.00	**50.67**
Mean	252.7		

Because of the squaring trick, adding up the last column does not give us zero. It gives us 50.67. Then dividing by the 3 games gives us 16.89. Taking the square root to get us back to a measure of yards rather than square yards, $\sqrt{16.89} = 4.11$ yards, which is the standard deviation!

$\textbf{Variance} = 50.67/3 = 16.89$

$\textbf{Standard deviation} = \sqrt{16.89} = 4.11$ yards

As we expected, a very small number compared to his 89.72 standard deviation in the preseason.

Empirical rule: 68/95/99.7

Now we can introduce the 68/95/99.7 rule (or **the empirical rule**) for standard deviation. This rule tells us that 68% of all our data values will lie within one standard deviation of the mean. (We just calculated one standard deviation, the 4.11 yards for Brady's regular-season passing.) The 95 tells us that 95% of all of our data lies within two standard deviations of the mean. And you can probably guess that 99.7% of the data lies within three standard deviations of the mean.

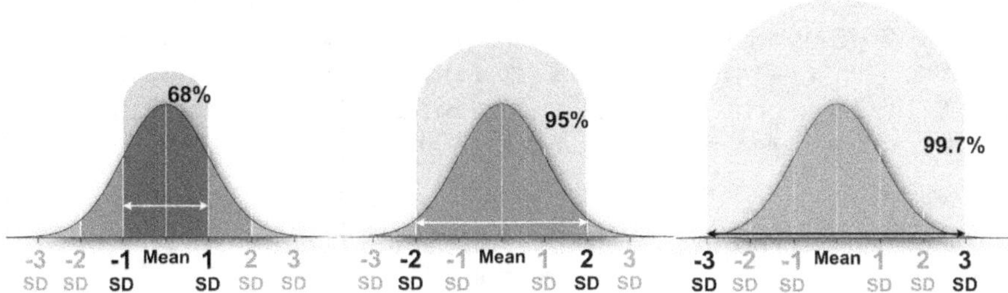

Let's do that with the regular-season Brady data. We'll round his numbers to make it easier to see. We'll make the mean 253 and the standard deviation 4. 68% of Brady's games, therefore, will be between 4 yards below the mean (253-4) and 4 yards above the mean (253+4). Between 249 and 257. 95% of his games will be between 2 standard deviations (4 × 2 = 8) below the mean and 2 standard deviations (again 4 × 2 = 8) above the mean. That is, between 245 and 261 (253 - 8 and 253 + 8). Finally, 99.7% of his games will be between 12 yards below and 12 yards above. Which is between 241 and 265 yards.

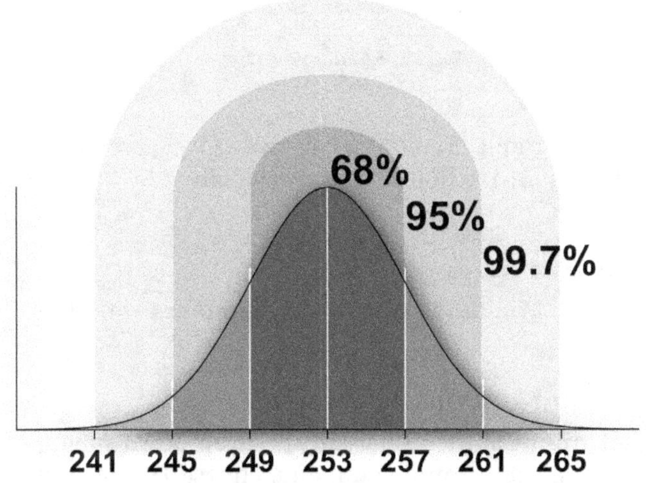

Tom Brady 2010 Passing Yards

This is how we use the concept of standard deviation in statistics.

It will be even clearer with more data. We ended up with 100% of Tom's data within 2 standard deviations. That's not unusual with so little data.

Let's do one more example from MLB. Here is the histogram for MLB in 2015, looking at batting average for all qualified hitters in both leagues. The N = 141 means we have 141 hitters to work with. Compare that to the three games we had for Tom Brady. The larger the n, the more comfortable we are with the descriptive statistics.

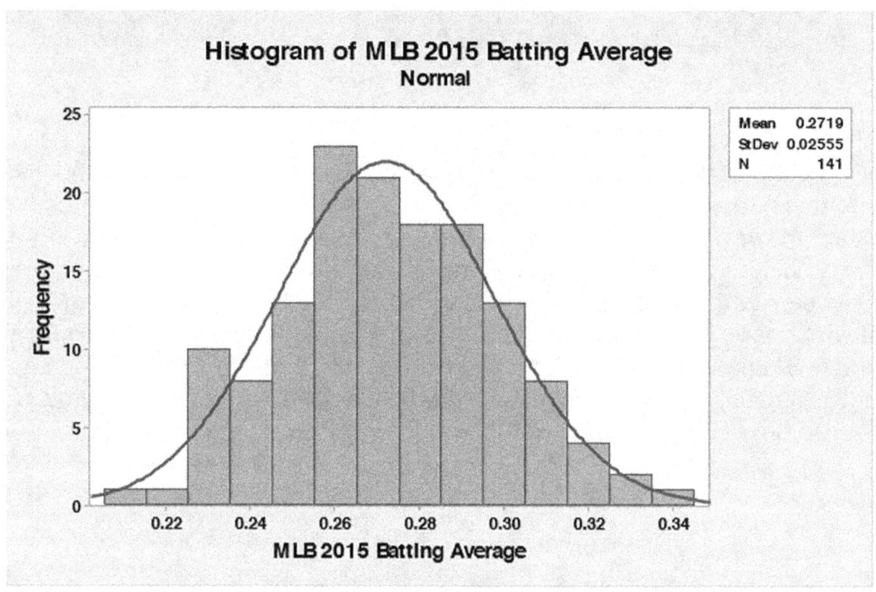

Notice the graph from Minitab also gives us the mean and the standard deviation! We don't have to calculate it by hand the way we did with Tom Brady.

Enter data in Minitab

We can also use Minitab to calculate those values without the graph. Put our data in C1.

Choose Minitab options

Choose Stat from the menu bar. In the drop-down menu, choose Basic Statistics and in that drop-down menu, choose Display descriptive stats. Double click on C1, which will move C1 to the variables box. Then click on stats. Make sure the mean and the standard deviation are checked. Then hit OK, OK.

Done!

What you will get is:

Descriptive Statistics: MLB 2015 Batting Average

```
Variable                          Mean      StDev
MLB 2015 Batting Average       0.27194    0.02555
```

Notice the result agrees with our graph! Now we can do the 68/95/99.7 rule for MLB hitters in 2015. The mean goes in the middle and then we add one standard deviation (SD) to the mean and subtract one standard deviation from the mean.

That shows us in 2015 68% of all MLB hitters hit between .246 and .297. 68% represents plus or minus one standard deviation from the mean. I rounded the mean to .272 and added the SD of .025, which gave me .297. Then I subtracted .025 from the mean, and that gave me .247. Within one SD of the mean just means adding it to the mean once, and subtracting it from the mean once.

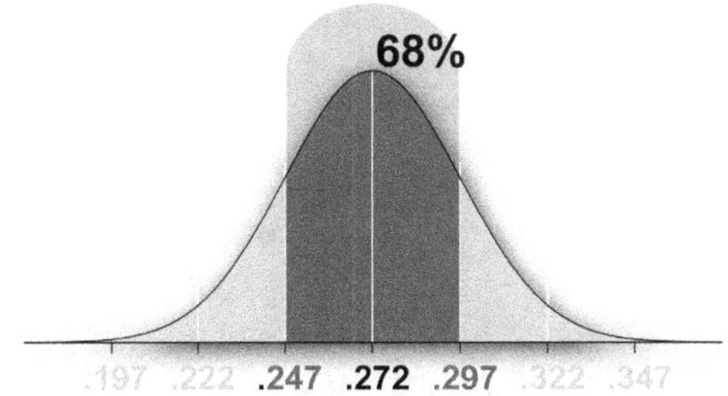

MLB 2015 Batting Averages
1 Standard Deviation

Now, when I add the same SD again (-.297 + .025), I get .322. And when I subtract the SD again (-.247 - .025), I get .222.

That means 95% of all MLB hitters in 2015 hit between .222 and .322, or plus and minus two standard deviations. Right? To get up to .322, I started at the mean (.272) and added .025 once to get to .297, then added it again to get to .322. And I subtracted it twice on the other side. 95% is always plus and minus 2 standard deviations from the mean.

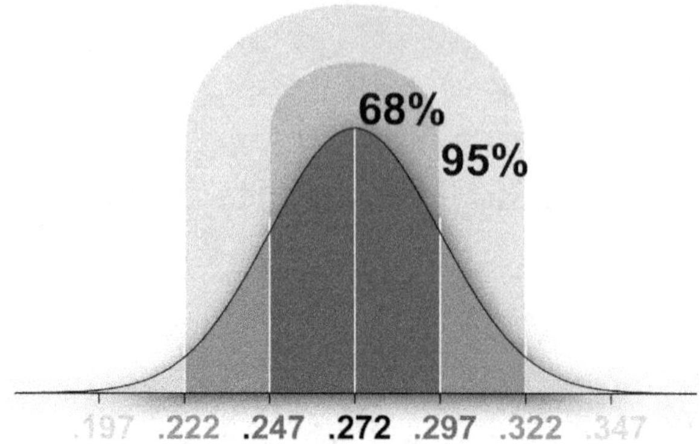

MLB 2015 Batting Averages
2 Standard Deviations

The largest is plus and minus three standard deviations, and that includes 99.7% of MLB hitters in 2015. Since that is just about every major league hitter, we can stop here. Add .025 one more time, and subtract it one more time. 99.7% of MLB hitters in 2015 hit between .197 and .347!

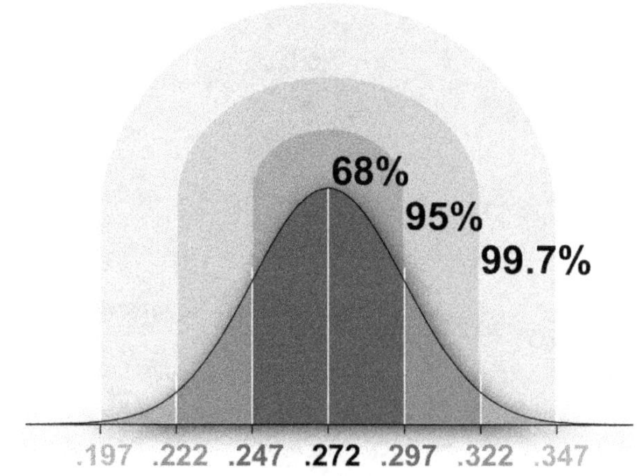

MLB 2015 Batting Averages
3 Standard Deviations

Hopefully, you feel a little more comfortable with the concept of standard deviations now. They just help us to see how far apart the data is, then we can apply the 68/95/99.7 rule to our data and interpret the result.

OPTIONAL: STANDARD DEVIATION BY HAND WITH THE PATRIOTS

The standard deviation measures the spread of symmetrical data, that is how far apart the values are. Specifically, how far the values (x) are from the mean ($\bar{x}$). It measures the standard difference, or the average difference, from the mean.

To calculate the difference is easy. We can do our values minus the mean. And then divide by the total number of values to get the average. But, because some values are negative and some positive, when we add them up, we will get zero. Now we do a math trick. We square those values to make them all positive, then we can divide by the total. That gives us a value called the variance. The reason it is not the standard deviation is we squared the differences earlier, now we have to un-square them (take the square root) to get back to our original units. The formula looks like this:

$$\text{Standard deviation for a sample} = \sqrt{\frac{\sum_{i=1}^{n}(x_i - \bar{x})^2}{n-1}}$$

$$\text{Standard deviation for a population} = \sqrt{\frac{\sum_{i=1}^{n}(x_i - \bar{x})^2}{n}}$$

A population is everything we're interested in looking at. The whole season. The whole career. A sample is data from a piece of the population. The first three games of a season. One year of a career.

If we have everything we use the *population* standard deviation formula. If we have a sample — a piece — of the data we use the *sample* standard deviation formula. One of the quirks of Minitab is it always calculates the standard deviation for a sample even when we have the whole population. So we'll use the sample formula so we can check how well Minitab does.

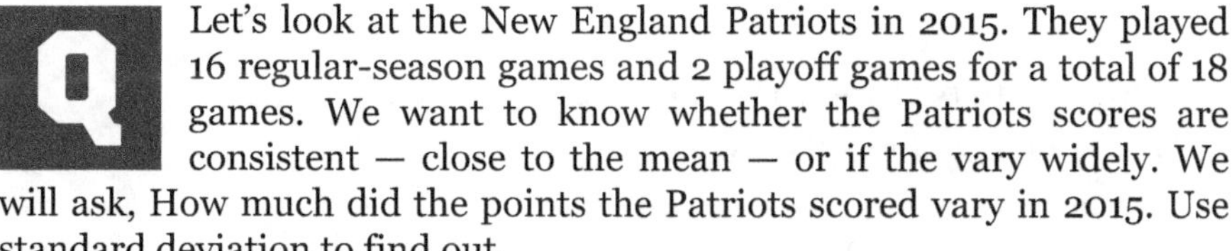

Let's look at the New England Patriots in 2015. They played 16 regular-season games and 2 playoff games for a total of 18 games. We want to know whether the Patriots scores are consistent — close to the mean — or if the vary widely. We will ask, How much did the points the Patriots scored vary in 2015. Use standard deviation to find out.

Here's the formula for standard deviation using a sample again:

$$\sqrt{\frac{\sum_{i=1}^{n}(x_i - \bar{x})^2}{n-1}}$$

The first easy part is n. That's the number of games the Pats played.

n - 1 = 17

The second easy part is the mean, x-bar. We add the points scored for each of the 18 games:

28+40+51+30+34+30+36+27+27+20+24+28+27+33+20+10+27+18=510

then divide the total by the number of games to get the Pats' mean, the average points per game.

$$\mathbf{x\text{-}bar}\text{ (the mean) } = \frac{\text{Pats's total points for 2015}}{\text{Number of games Pats played in 2015}}$$

$$= \frac{510}{18} = 28.33$$

Now we get to a part that may be less familiar to many. In the standard deviation formula there is the Greek letter sigma (Σ). That's math shorthand for, "Do the same calculation on each piece of data then add them all up." In our example, that means take the points scored in the first game, subtract the mean, then square it. Then do that for the second game. And for the third game. And so on until we've done it for all 18 games. *And then* add all those up.

If we had to write it all out as a formula, it would look like:

(Game 1 points - Average points)^2

+ (Game 2 points - Average points)^2

+ (Game 3 points - Average points)^2

...

+ (Game 16 points - Average points)^2

+ (Game 17 points - Average points)^2

+ (Game 18 points - Average points)^2

18 is bad enough. If we were working with a baseball season, we'd need to do that 162 times! Instead, for the Pats, we write it like:

$$\sum_{i=1}^{18} (\text{Game } i \text{ points} - \text{Average points})^2$$

Much simpler!

Simpler to write but still not easy to subtract then square each of the 18 game points then add them all up at the end. When I took statistics in college, that's exactly what I had to do. I won't ask you to do that. I won't do it! I'll cheat on my "by hand" and use Minitab. (If you're clever with a spreadsheet, you could also use that.)

SUM ($\sum$) of each of the squared game points minus the mean.

$$(28 - 28.33)^2 + (40 - 28.33)^2 + \dots + (18-28.33)^2 = 1376$$

Now we'll plug numbers into the formula.

$$\sqrt{\frac{\sum_{i=1}^{n} (x_i - \bar{x})^2}{n - 1}}$$

$$= \sqrt{\frac{\text{SUM}}{\text{number of games - 1}}}$$

Plug in: $\qquad = \sqrt{\dfrac{1376}{17}}$

Divide by n-1: $= \sqrt{80.94}$ **Variance** (under the radical)

Square root: $= 9.00$ **Standard deviation**

And we're done!

Let's double check on Minitab.

Enter data in Minitab

I entered the points the Pats scored for each of the 18 games into C1 in Minitab.

Choose Minitab options

Choose Stat from the menu bar, Basic statistics, Display descriptive stats, and then clicked on statistics and checked standard deviation.

Done!

Same answers. Much faster!

Descriptive Statistics: ...

WORKSHEET 1

Descriptive Statistics: Pats points scored 2015

Statistics

Variable	Total Count	Mean	StDev	Variance	Sum
Pats points scored 2015	18	28.33	9.00	80.94	510.00

	C1	C2	C3	C4	C5
	Pats points scored 2015				
1	28	*	*		
2	40				
3	51				
4	30				

CHAPTER 13

Z-scores

The only important statistic is the final score.

— Bill Russell, Hall of Fame Celtics legend

PREVIEW: WHO HAD THE BETTER DAY?

Now we can use standard deviation to help us calculate another statistic, namely a z-score.

A **z-score** tells us how many standard deviations we are from the mean. A z-score of one means we are one standard deviation above the mean. A score of -1 means we are 1 standard deviation below the mean. A score of 1.6 means we are 1.6 standard deviations above the mean. A score of -.8 means we are .8 standard deviations below the mean. A z-score of zero means we are at the mean. When we calculate a z-score we are calculating how far something is from the mean. Keep in mind a negative z-score is below the mean, a positive z-score is above the mean.

Those facts are all well and good, but is the z-score formula too complicated? No! It looks like this:

$$\text{z-score} = \frac{X - \bar{X}}{SD}$$

We take the value we are looking at and call it x. Then subtract the mean (x-bar). Then divide that answer by the standard deviation. The result is our z-score. Is it that easy? Yes!

Z-scores will help us answer the question of who performed better when we have two athletes in two different sports or different positions. For instance, who had a better year in 2015 in the NFL, the passing yards leader for quarterbacks (Drew Brees), or the running-back king (Adrian Peterson)? While they both were the statistical leaders for their positions, their positions are entirely different. How can we compare them? With a z-score. The z-score will standardize their results so we can compare them. That is, the z-score will tell us how far each is from the average for their position. We don't need to guess how far they stand out. A z-score will tell us exactly.

For instance, one standard deviation is better than one half, right? And two standard deviations are better than one. This is because you are further from an average (mean) performance. If we standardize both values (passing yards and rushing yards), we can compare them apples to apples and see who did better!

By the way, measuring who did better means we have to forget about the positives and negatives. We just compare the distance from the mean in either direction. Whoever is farther from the mean had the more surprising result, or the better performance.

MIGUEL CABRERA TEACHES US HOW TO FIND A Z-SCORE BY HAND

Miguel Cabrera was the AL Batting champ in 2015, with a .338 average. How far was his average from the average of all batters in 2015? In statistics language, What was Cabrera's z-score?

To save time, we'll let Minitab do the number crunching to get the AL batting average for all qualified players in 2015 and their standard deviation.

Cabrera's average = .338

AL average (Mean) = .2719

2015 batters' SD = .02546.

To get the z-score we do:

$$\text{z-score} = \frac{X - \bar{X}}{SD} = \frac{\text{Cabrera's average - AL average}}{\text{2015 batters' SD}}$$

Plug in	$= \dfrac{.338 - .2719}{.02546}$
Subtract	$= \dfrac{\mathbf{.0661}}{.02546}$
Divide	$= \mathbf{2.60}$

We are done! Miguel Cabrera had a **z-score of 2.60** in 2015.

That means Miguel Cabrera was 2.6 standard deviations above the mean in 2015, which is very high! In statistics, we would consider it unusual because it is more than 2 standard deviations away from the mean.

GUIDED EXAMPLES FOR Z-SCORES WITH SC AND TED WILLIAMS

Q1 Kellie Pennington of Springfield College was first at the 2012 NCAA Division III 50 Yard Freestyle National Championship in 23 flat. At the same time, Brian Fuller of Springfield College finished third at the NCAA Division III National Championships in Track and Field in the Steeplechase. They were both All-Americans and had phenomenal performances. Yet who had the better day?

Women's 50 Yard Freestyle

Pl	Name	College	Time
1	Pennington, Kellie	Springfield College	23:00
2	Raleigh, Christie	Rowan Univ	23:09
3	Pavlak, Claire	Emory Univ	23:10
4	Pierce, Laura	The College of NJ	23:31
5	Rosenkranz, Renee	Emory Univ	23:35
6	Dobben, Anna	Emory Univ	23:47
7	Nuess, Morgan	Denison Univ	23:49
8	Lukes, Chandra	Univ of Redlands	23:82

Men's 3000M Steeplechase

Pl	Name	Year	School	Finals	Points
1	Nick Kramer	SR	Calvin	8:51.01	10
2	Jack Davies	JR	Middlebury	8:54.61	8
3	Brian Fuller	SR	Springfield	8:56.05	6
4	Michael LeDuc	SO	Conn College	8:59.54	5
5	Anders Crabo	SR	Pomona-Pitzer	9:00.89	4
6	Trevor Siperek	SR	Coast Guard	9:01.53	3
7	Bobby Over	JR	Allegheny	9:03.00	2
8	Jared Brandenburg	SR	Wis.-River Falls	9:04.69	1
9	Stephen Serene	SR	MIT	9:09.23	
10	Andrew Wortham	JR	Bates	9:19.40	
11	Brandon Abasolo	JR	Williams	9:20.78	
12	Marcus Huderle	SO	Carleton	9:22.21	
13	Tyler Morey	SR	Wis.-Oshkosh	9:22.38	
14	Jeremy Kieser	JR	Wis.-Eau Claire	9:34.80	

How can we compare a finisher in the 3000M steeplechase for men with a finisher in the 50M freestyle swim for women? They are not just different distances and different genders, but different sports.

We can do that by first converting the data to one standard measure. That measure? Yes, of course, the z-score. We'll ask, How many standard deviations was each from their competition's mean? How much better were they than everyone else they competed against?

Enter data in Minitab

To calculate the z-score on Minitab, go to C1 and enter the 50M Freestyle swim times. In C2, enter the 300M Steeplechase run times. Notice that we enter them in seconds so the program knows that they are quantitative data. Enter your titles in C1 and C2, as in the example below. Then label C3 Swim Z-Score and label C4 Run Z-Score. Minitab will use those columns to store the results (the actual z-scores) it calculates.

Choose Minitab options

Choose Calc from the menu bar and pick Standardize. Put your cursor in the box for the input column, then double click on C1 and C2. Go to the Store results in box and double click on C3 and C4. Click OK. *You* are done! Now it's Minitab's turn. Minitab calculates the z-score for us and lists them in C3 and C4.

Worksheet 1 ***			
	C1	C2	C3	C4
↓	50m Freestyle	3000m Steeplechase	Swim Z score	Run Z score
1	23.31	531.01	-0.06977	-1.34218
2	23.49	534.61	0.60006	-1.06718
3	23.00	536.05	-1.22337	-0.95717
4	23.10	539.54	-0.85125	-0.69057
5	23.09	540.89	-0.88846	-0.58744
6	23.35	541.53	0.07908	-0.53855
7	23.47	543.00	0.52563	-0.42626
8	23.82	544.69	1.82808	-0.29716
9		549.23		0.04965
10		559.40		0.82655
11		560.78		0.93197
12		562.21		1.04120
13		562.38		1.05419
14		574.80		2.00296

Done! Now we can compare Kellie's performance to Brian's. Notice Kellie has a z-score of -1.22337. That means she was 1.22 standard deviations below the mean for that race. She was a bit more than one standard deviation faster than the average. Then look at Brian. He had a -.95717 z-score. His performance was a bit less than one standard deviation below the mean. He wasn't quite as far from the mean as Kellie was. Therefore, Kellie had the better day. She was farther from the mean. The further you are from the mean, the better you are.

Compare Ted Williams's batting average in 1941 (his .406 that year made him the last major league player to break .400!) with Tom Brady's passing yards in 2011 (his 5235 yards in 2011 was a career-best) using z-scores and answer the question, Which player was further away from the average? In other words, which had the better season?

I used Minitab to calculate the mean and standard deviation for us to save us the tedium. Note, for Ted I used only qualified hitters (at least 3.1 plate appearances per game) and for Tom I used only players with the three-game minimum. Minitab tells us:

Ted Williams:

$$= \frac{\text{Ted's average - 1946 MLB average}}{\text{1946 batters' SD}}$$

$$= \frac{.406 - .28646}{.03704}$$

z-score = 3.23

Tom Brady:

$$= \frac{\text{Tom's average - 2011 NFL average}}{\text{2011 players' SD}}$$

$$= \frac{5235 - 2568}{1468}$$

z-score = 1.82

Ted Williams had the better season because his z-score of 3.23 is farther from 0 compared to Tom Brady's 1.82

To see how Minitab handles z-scores, **Enter data in Minitab.** Put the 1946 MLB batting averages into C1 and the 2011 NFL passing yards into C2. Label C3 MLB z-scores and label C4 NFL z-scores. Minitab will calculate the z-scores and put them in those columns.

Choose Minitab options. On Minitab, go to Calc in the menu, then choose Standardize (remember, a z-score is a standardized score). Double click on C1 and C2 to tell Minitab to use those columns as the input columns, then double click on C3 and C4 to tell Minitab where to Store the results in.

That's it! Notice it defaults to subtract mean and divide by the standard deviation, which is what you did by hand above. Then click OK, and it will fill in the z-scores for both columns.

With both MLB and NFL numbers converted to a standard value, we can compare them. It's now fair to ask whether Ted Williams or Tom Brady had the better season and be comfortable saying that Ted Williams had the better season. Not only better, but incredible! Remember we remarked how high Miguel Cabrera's 2.6 z-score was? Ted Williams is over 3 standard deviations from the mean!

Worksheet 2 ***

↓	C1	C2	C3	C4
	1946 MLB Batting Averages	2011 NFL Passing Yards	MLB z-scores	NFL z-scores
1	0.406	5038	3.22712	1.68243
2	0.359	5476	1.95832	1.98081
3	0.357	5235	1.90433	1.81663
4	0.340	4933	1.44540	1.61090
5	0.334	4624	1.28343	1.40039
6	0.324	3832	1.01347	0.86084
7	0.322	4177	0.95948	1.09587
8	0.322	3592	0.95948	0.69734
9	0.317	3474	0.82450	0.61695
10	0.314	3610	0.74351	0.70960
11	0.311	4184	0.66252	1.10064
12	0.311	3571	0.66252	0.68303
13	0.307	4051	0.55454	1.01003
14	0.301	3398	0.39256	0.56518
15	0.300	4077	0.36557	1.02775
16	0.299	4643	0.33857	1.41333
17	0.298	2733	0.31158	0.11215
18	0.298	3151	0.31158	0.39691
19	0.297	3091	0.28458	0.35603
20	0.297	3144	0.28458	0.39214
21	0.288	3303	0.04162	0.50046
22	0.284	2214	-0.06636	-0.24142
23	0.283	2164	-0.09336	-0.27549
24	0.282	2497	-0.12036	-0.04863
25	0.280	2753	-0.17435	0.12577
26	0.279	2319	-0.20134	-0.16989
27	0.277	2479	-0.25534	-0.06089
28	0.276	1853	-0.28233	-0.48735
29	0.275	1913	-0.30933	-0.44648

Comparing negatives and positives

Sometimes one athlete's z-score can be negative while the other athlete's z-score is positive. In track, runners want go faster than the mean (below the mean, so negative) and throwers want more distance (above the mean, so positive). Do we evaluate them differently? No. Which z-score is further from zero (the mean) had the better day.

Ted Williams has a positive z-score (3.23) for hitting .406 in 1941. Geoffrey Mutai has a negative z-score (-2.78) compared to the first 100 finishers in winning the Boston Marathon in 2011. One positive, one negative, but who had the better result?

But, some students cringe when comparing negatives and positives, so there is an easy trick to fix that. Use absolute values. Since the absolute value is always a positive number, you can compare them apples to apples.

The absolute value for Williams is still a positive 3.23, and for Mutai a positive 2.78. 3.23 is bigger than 2.78. Easier now to see Williams had the better result because his z-score of 3.23 is further from zero than Mutai's -2.78

For instance, comparing Williams' 3.23 and Brady's 1.82 is easy. 3.23 is further from zero, so 3.23 had the better year. No need for absolute values.

But how about comparing Mutai's -2.78 and Miguel Cabrera's 2.60? If you can see that -2.78 is further away from 0 than 2.60 is, well you are right! Do not change anything. -2.78 had the better performance.

BUT, if that was not evident to you, take the absolute value first. The absolute value of -2.78 is 2.78. When you compare 2.78 and 2.60 you can see that 2.78 is further from zero, so they had the better result. But remember when you answer the question on the exam, tell me the -2.78 had the better result because it was further from zero than 2.60.

Absolute values also work if both numbers are negative. Comparing -1.44 and -1.49 is sometimes not easy. But taking the absolute value and comparing 1.44 and 1.49? Much easier! 1.49 is bigger. 1.49 is further from 0. But remember when you write your answer, switch back to the negative value. The answer will be -1.49 because -1.49 is further from zero than -1.44.

OPTIONAL: ANOTHER LEVEL OF Z-SCORE WITH PHELPS AND BOLT

You now know everything you need to about z-scores for the exam. On the exam you will always compare two athletes' performances. It might be a single event for each. Or a whole season. Or a whole career.

But Phelps and Bolt show us that we can take z-scores to another level. We can compare two athletes who have competed in multiple events. For instance, we have three different swimming events from the 2016 Olympics for Phelps and two different track events from the same Olympics for Bolt.

Which athlete had the better 2016 Olympics in individual (non-relay) races, Michael Phelps or Usain Bolt? Who was better compared to the rest of their competitors?

In the Minitab output below are the 2016 Olympic-finals times for the 3 individual races of Michael Phelps and 2 of Usain Bolt. This allowed Minitab to calculate five separate z-scores, one for each race.

Phelps's results are in blue (C1-C3). Bolt's in orange (C4-C5).

Remember, Minitab does three steps to calculate a z-score. The first step is to calculate the mean (average) time for each race. In its memory, Minitab has the mean time for five different races.

In Step 2 Minitab takes the athlete's value (his time) and subtracts the mean for that race. Then in Step 3 Minitab divides by the standard deviation for that race. That gives the z-score for each finisher. Minitab does that calculation for us and loads the z-scores into C6-C8 (Michael Phelps) and C9-C10 (Usain Bolt).

For example, Phelps's z-score for the 200 m Individual Medley is -2.13. That means his time is more than 2 standard deviations below the mean. That's huge. No one else comes close. And negative is a good thing in this case since, for swimmers, lower times are better.

But Phelps only had a z-score of -1.01 in the 200M Butterfly. In the 100M Butterfly he finished second and had -.3. Adding the z-scores and dividing by 3 gives an average z-score of -1.15.

Usain Bolt had z-scores of -1.78 and -1.65. Those average to -1.72, much lower than Phelps's average.

According to the z-scores, even though Phelps had the best race, Bolt had the better Olympics performance because his total performance was better than Phelps.

PRACTICE PROBLEM

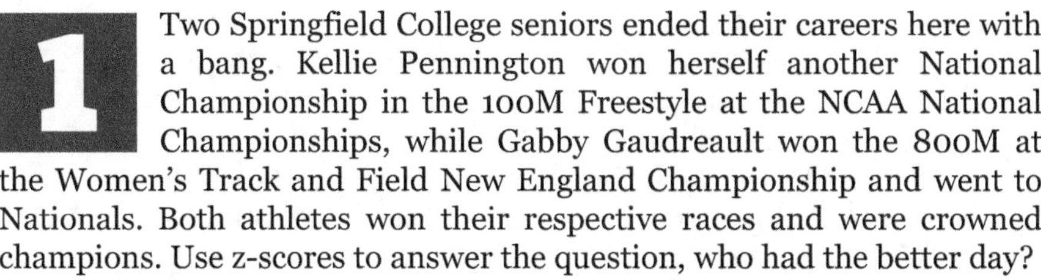

Two Springfield College seniors ended their careers here with a bang. Kellie Pennington won herself another National Championship in the 100M Freestyle at the NCAA National Championships, while Gabby Gaudreault won the 800M at the Women's Track and Field New England Championship and went to Nationals. Both athletes won their respective races and were crowned champions. Use z-scores to answer the question, who had the better day?

Notice I copied the results for both races below. I did not answer the question for you. You will practice finding the answer on Minitab. But I will give you a hint. You will need just four columns in Minitab total to answer this question. On the input side you only need the athlete's times in seconds for C1 and C2.

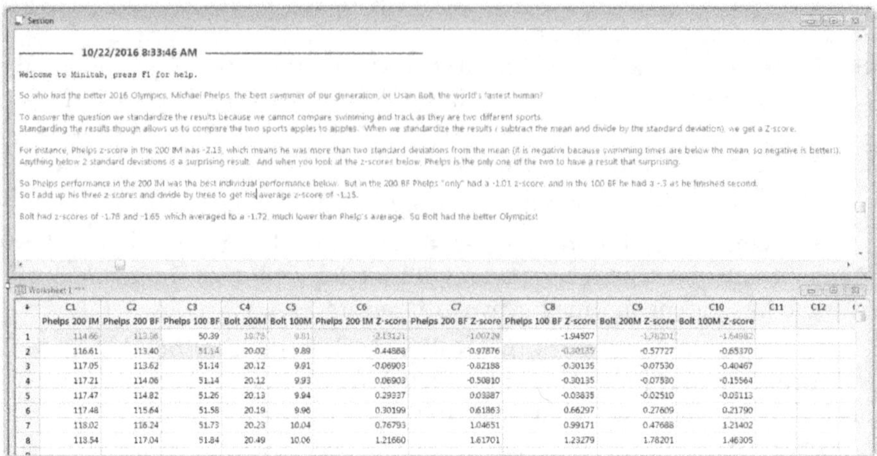

Session

———— 10/22/2016 8:33:46 AM ————

Welcome to Minitab, press F1 for help.

So who had the better 2016 Olympics, Michael Phelps, the best swimmer of our generation, or Usain Bolt, the world's fastest human?

To answer the question we standardize the results because we cannot compare swimming and track as they are two different sports.
Standarding the results though allows us to compare the two sports apples to apples. When we standardize the results (subtract the mean and divide by the standard deviation) we get a Z-score.

For instance, Phelps z-score in the 200 IM was -2.13, which means he was more than two standard deviations from the mean (it is negative because swimming times are below the mean so negative is better). Anything below 2 standard deviations is a surprising result. And when you look at the z-scores below, Phelps is the only one of the two to have a result that surprising.

So Phelps performance in the 200 IM was the best individual performance below. But in the 200 BF Phelps "only" had a -1.01 z-score, and in the 100 BF he had a -.3 as he finished second.
So I add up his three z-scores and divide by three to get his average z-score of -1.15.

Bolt had z-scores of -1.78 and -1.65 which averaged to a -1.72, much lower than Phelp's average. So Bolt had the better Olympics!

Worksheet 1 ***

	C1	C2	C3	C4	C5	C6	C7	C8	C9	C10	C11	C12
	Phelps 200 IM	Phelps 200 BF	Phelps 100 BF	Bolt 200M	Bolt 100M	Phelps 200 IM Z-score	Phelps 200 BF Z-score	Phelps 100 BF Z-score	Bolt 200M Z-score	Bolt 100M Z-score		
1	114.66	113.56	50.39	19.78	9.81	-2.13121	-1.00729	-1.94507	-1.78201	-1.64982		
2	116.61	113.40	51.14	20.02	9.89	-0.44888	-0.97876	-0.30135	-0.57727	-0.65370		
3	117.05	113.62	51.14	20.12	9.91	-0.06903	-0.82188	-0.30135	-0.07530	-0.40467		
4	117.21	114.06	51.14	20.12	9.93	0.06903	-0.50810	-0.30135	-0.07530	-0.15564		
5	117.47	114.82	51.26	20.13	9.94	0.29337	0.03887	-0.03835	-0.02510	-0.03113		
6	117.48	115.64	51.58	20.19	9.96	0.30199	0.61863	0.66297	0.27609	0.21790		
7	118.02	116.24	51.73	20.23	10.04	0.76793	1.04651	0.99171	0.47688	1.21402		
8	118.54	117.04	51.84	20.49	10.06	1.21660	1.61701	1.23279	1.78201	1.46305		

Springfield, MA - May 3, 2014 - SC's Gabriella Gaudreault won the 800 m championship on Saturday at the NE Division III Track & Field Championships.

	Name	Class	College	Time
1	Gabriella Gaudreault	SR	Springfield	2:13.18(133.18)
2	Sarah Fusco	JR	Bates	2:14.48(134.48)
3	Cindy Huang	JR	MIT	2:15.44(135.44)
4	Hannah Damron	SO	S. Maine	2:15.66(135.66)
5	Lauren Gormer	JR	Tufts	2:16.58(136.58)
6	Marina Capalbo	JR	Babson	2:17.17(137.17)
7	Chrissy Larrabee	JR	Husson	2:17.24(137.24)
8	Sydney Smith	SO	Tufts	2:17.68(137.68)
9	Sharon Ng	SO	Wellesley	2:18.21(138.21)
10	Nikita Rajgopal	FR	Wesleyan	2:20.05(140.05)
11	Julie Tevenan	SO	WPI	2:20.28(140.28)
12	Ashley Monahan	FR	Westfield St	2:21.32(141.32)
12	Samantha Pomroy	SO	Springfield	2:21.32(141.32)
14	Keelin Moehl	SO	Amherst	2:21.82(141.82)
15	Louise van den Heuvel	SR	MIT	2:21.90(141.90)
16	Addis Fouche-Channer	FR	Middlebury	2:21.92(141.92)
17	Olivia Tarantino	JR	Amherst	2:23.84(143.84)

Indianapolis, IN. – March 22, 2014 - Senior Kellie Pennington (Monson, MA) put the exclamation point on her historic career by winning the NCAA Division III National Championship in the 100 freestyle on Saturday night.

	Name	Class	College	Time
1	Pennington, Kellie	SR	Springfield	49.41
2	Bogdanovski, A.	JR	Johns Hopkins	49.66
3	Larson, Nancy	JR	Emory	50.33
4	Yarosh, Hillary	SR	Kenyon	50.94
5	Cline, Jourdan	JR	Kenyon	50.95
6	Pielock, Julia	SR	Connecticut	51.02
7	Kane, Carolyn	FR	Denison	51.23
8	Ternes, Kylie	SR	Johns Hopkins	51.33

Practice exams for Part 1

On the next few pages, you will find copies of previous Exam 1s given at Springfield College. Feel free to use them for practice.

Good luck!

SPRINGFIELD COLLEGE
Sports Statistics  Exam 1 Spring 2020

Springfield College Women's basketball is also having a good year, as they are 8-1 in NEWMAC play and they give SC a good chance at having both women's and men's teams playing in March Madness in 2020.

In question 1, we will examine their roster by class to see if Coach Graves is set-up for continued success, or if she has issues with the team. To answer this question, create a graph of the SC women's basketball team by class (CL).

 Using the same data from Question 1, perform a chi-square analysis on the women's roster by class to measure how well the roster fits your expected numbers. Identify the six components of the hypothesis test in your answer:

Hypothesis Test

1. **H_0:**
2. **H_A:**
3. **Test statistic and its value:**
4. **p-value:**
5. **Statistical conclusion:**
6. **Interpretation:**

 The Boston Bruins are the NHL's best team in the 2019-20 season with 88 points, overcoming a difficult ending last season (losing Game 7 at home in the finals when they were favored). We will give some love to the Bruins by Googling "Bruins Stats" and clicking on the NHL.com link. Copy the fourth column for points (P) into Minitab for all players except the two goalies.

Create an appropriate graph for points for the almost-Stanley-Cup-champions Boston Bruins.

Q4 The Springfield College football team had the number one rushing offense in all of Division III in 2019. Yet cross-town rival WNEU went to the NCAAs. Which team had the better rushing totals last fall?

Use back-to-back boxplots to answer the question of which Springfield school had the better rushing offense in 2019?

SC rushing	WNEU rushing
312	172
420	177
171	161
305	313
404	132
277	223
338	199
470	188
561	207
316	210

Q5 We just saw that David Pastrnak of the Boston Bruins is leading the NHL in goals scored. He, therefore, is leading the Bruins as well. You already loaded the Bruins data into Minitab, so we will use that same column here for Question 5.

Jake Ross is leading the Springfield College basketball team in scoring. He is 8th in the country in all of Division III with 25.2 points per game. Enter the average points per game for each of the SC players into Minitab:

Name	Avg PPG
Trey Witter	10.2
Deonte Sandifer	2.6
Kendall Baldwin	1.7
Collin Lindsay	8.0
Jake Ross	25.2
Jake Jacobson	1.4
Daryl Costa	6.5
Charlie Clay	4.4
Robert Baum	5.9
Noah Cummings	2.5
Heath Post	17.7
Joe Fontanella	2.0
Harper Niven	1.9
Panayiotis Kapanides	0.8

Which player is having the better season? David Pastrnak in points scored for the Bruins? Or Jake Ross in average points per game scored for SC? Use z-scores to answer the question and include the actual z-score in your answer.

SPRINGFIELD COLLEGE
Sports Statistics  Exam 1 Fall 2019

 Six Springfield College athletes qualified for the NCAA Division III outdoor track and field championships this year. The meet was held on May 23-24 in Geneva, Ohio. Two of those athletes earned All-American status, Caroline Hitchcock and Carley Moyher.

Caroline finished 5th in the high jump, and Carley finished 8th in the Javelin. Both were named to the All-American team, but which had the better day? Use z-scores to answer the question and include the actual z score in your answer.

Event 25 Women High Jump

Place	Name	Year	School	Final
1	Cirrus Robinson	JR	Ohio Wesleyan	1.70m
2	Laura Pullins	SO	Marietta	1.70m
2	Laura Darcey	JR	U. of Chicago	1.70m
4	Emma Egan	JR	Williams	1.70m
5	Caroline Hitchcock	SO	Springfield	1.70m
6	Lauren Wilson	SO	Pacific Lutheran	1.70m
6	Jacy Scharlow	FR	RPI	1.70m
8	Lauren Parker	SR	RPI	1.67m
9	Grace Bordson	SR	St. Thomas (Minn.)	1.67m
10	Sarah Gomez	FR	Carroll (Wis.)	1.64m
11	Summer-Solstice Thomas	JR	Williams	1.64m
12	Kennady Gibbins	FR	Mount Union	1.64m
13	Lizz Ottusch	SR	Wis.-Eau Claire	1.64m
14	Reilly Wagner	SR	Misericordia	1.61m
14	Taylor Wiederrecht	JR	Messiah	1.61m
16	Bailey Waldhauser	SO	Wis.-Eau Claire	1.61m
16	Kassidy Mulryne	FR	TCNJ	1.61m
16	Rachel Ohde	SO	Saint Mary's	1.61m
16	Sara Kate Capel	JR	Rhodes	1.61m
20	Stephanie Pladies	SR	Wis.-Stout	1.61m
21	Michaela Chowning	FR	Eastern Mennonite	1.61m
22	Isabella Bruno	SO	Trinity (Conn.)	1.56m

Event 39 Women Javelin

Place	Name	Year	School	Final
1	Brooke France	FR	Pacific (Ore.)	44.64m
2	Veronica Montane	SO	Johns Hopkins	44.45m
3	Sophia Slovenski	FR	Bowdoin	44.13m
4	Alex Grubbs	JR	Wis.-Stevens Point	43.97m
5	Katy McClellan	SR	Smith College	42.23m
6	Britney Carnagie	SO	Adrian	42.11m
7	Ava Nelson	FR	Pacific Lutheran	41.31m
8	Carley Moyher	JR	Springfield	41.20m
9	Kathleen Merchant	FR	St. Lawrence	40.94m
10	Devan Walsh	JR	Trinity (Conn.)	40.59m
11	Cathryn MacGregor	SR	Wheaton (Mass.)	40.27m
12	Kylie Orndorf	SR	Juniata	39.84m
13	Emily Sproul	SR	Pacific (Ore.)	39.52m
14	Chrissy Strickland	FR	George Fox	39.51m
15	Jensyn Lown	SO	George Fox	38.94m
16	Savannah Freeman	SR	McMurry	38.87m
17	Cayle Spencer	JR	King's (Pa.)	38.00m
18	Emily Smatlak	SO	Anderson (Ind.)	37.45m
19	Payton Passantino	JR	Millsaps	37.23m
20	Samantha Liberty	SR	Springfield	36.11m
21	Tia Hart	FR	Texas Lutheran	35.71m
22	Rebecca Gorman	JR	Middlebury	33.51m

Q2 The Springfield College Women's volleyball team lost a close contest to Coast Guard last night, falling to 12-5 on the year. The loss happened despite the efforts of superstars Anagabrielle Sanchez and Camryn Bancroft. They are on pace to do well in the postseason once again.

But are they set up for continued success? We will examine their roster by class and create an appropriate graph for that data.

Q3 Using the same data from Question 2, perform a chi-square analysis on the women's volleyball roster by class to measure how well the roster fits your expected numbers. Identify the six components of the hypothesis test in your answer.

Hypothesis Test

1. **H_0:**
2. **H_A:**
3. **Test statistic and its value:**
4. ***p*-value:**
5. **Statistical conclusion:**
6. **Interpretation:**

Tomorrow the Bruins open their 2019 season on the road against the Dallas Stars. This will be the Bruins' first real game since the devastating, painful, gut-wrenching loss to the Blues at home in Game 7 of the Stanley Cup finals. How will they respond to the physical play of the long postseason and the emotional letdown of a Game 7 home loss? We will start to find out tomorrow!

For now, Google "Bruins Stats" and click on the NHL.com link. Scroll down and copy the Points column for all players into Minitab. Now create an appropriate graph for Points for the 2018-19 Boston Bruins.

The New England Patriots won the Super Bowl with an incredible defensive effort, holding the Rams to 3 points. Since then, they have been on a roll, giving up just their first touchdown last week to the Bills. But is the Patriots defense even better now than last year's Super Bowl-winning team?

Google "Patriots stats" and click on pro-football-reference.com. Scroll down to Schedule and Game results. Under Score copy the four games for this year under Opp (opponent) and paste them into Minitab. Then scroll back up and click Previous season and do the same for 2018 (include playoffs).

Use back to back boxplots to answer the question, Which Patriots defense is better, 2018 or 2019? Remember, better means fewer points allowed, so use Q1 instead of Q3 in your answer.

SPRINGFIELD COLLEGE
Sports Statistics  Exam 1 Spring 2019

 On Saturday, Chloe Dewhurst (Lancaster, MA) broke the 35-year-old high jump record for the SC track team, jumping 1.72 meters during the Triangle Classic. Also highlighting the day for the Pride was junior Katherine Strain (Longmeadow, Mass), who had an outstanding showing in the 60-meter hurdles as she sprinted to a 9.46 first-place finish. They both won their events, but who had the better day? Use z-scores to answer the question and include the actual z-scores in your answer.

The results for both athletes are below:

Women High Jump

Place	Name	Year	School	Final
1	Chloe Dewhurst	FR	Springfield	1.72m
2	Margaret Redfield	SO	MIT	1.60m
3	Mia Facchini	JR	Springfield	1.60m
4	Liana Reilly	SO	MIT	1.55m
5	Katherine Strain	JR	Springfield	1.55m
6	Hannah Miller	FR	Springfield	1.50m
6	Hannah Hu	FR	Springfield	1.50m
8	Isabella Bruno	SO	Trinity	1.50m
9	Paige Fielding	SR	Tufts	1.45m
10	Ashley Holton	FR	MIT	1.40m

Women 60 Meter Hurdles

Place	Name	Year	School	Final
1	Katherine Strain	JR	Springfield	9.46
2	Eizabeth Herlihy	FR	Springfield	9.72
3	Shannon Kennedy	SR	Connecticut	9.74
4	Mary Sims	JR	Coast Guard	9.90
5	Katelynn Lane	FR	Coast Guard	10.00
6	Ashley Consolini	JR	Springfield	10.00
7	Abigail Tata	FR	Springfield	10.38
8	Danielle Wood	SO	Springfield	10.44

 The Springfield College Women's Indoor Track and Field team is loaded with talent this year, and the women are looking to leave their mark this weekend up in Brunswick Maine at the Division III New England Championships, hosted by Bowdoin College.

We can examine their roster to see if Mike is set-up for continued success, or if he has issues with the team. To do this, create a graph of the women's indoor track roster by class (CL)

 Using the same data as above (problem 2), perform a chi-square analysis on the women's roster by class to measure how well the roster fits your expected numbers. Identify the six components of the hypothesis test in your answer:

Hypothesis Test

1. **H_0:**
2. **H_A:**
3. **Test statistic and its value:**
4. **p-value:**
5. **Statistical conclusion:**
6. **Interpretation:**

 The New England Patriots had their easiest Super Bowl victory (at least in terms of margin of victory) over the Rams this year. But the defensive battle was a surprise. The Rams had been the best offense in the NFC. Now that the season is over, let's compare. Who had the better offense this season? Google "Patriots stats" and go to pro-football-reference.com. Copy the total yards column. Then do the same for the Rams.

Paste the data into Minitab and use back-to-back boxplots to answer the question of which NFL team had the better offense in 2018-19?

The World Champion Red Sox have finally reported to spring training in Florida. Let's take one last look at the best team in baseball history, the 2018 World Champion Boston Red Sox. Google "Red Sox stats," click on mlb.com, then click on pitching. Copy the entire ERA column into Minitab then create an appropriate graph for ERA for the World Champs (did I mention they were World Champs?)

PART 2
Predictive Statistics

There are two kinds of statistics, the kind you look up and the kind you make up.

— Rex Stout, mystery writer

PREDICTION PERFECTED

CHAPTER 14

Simpson's paradox

> This isn't a paradox, and it wasn't actually discovered by Simpson, but that's the name everybody uses for it.
>
> — *Bentarm, author*

PREVIEW: SOMETHING THAT HAPPENS IN SPORTS EVEN THOUGH IT SHOULDN'T!

Karl Pearson first discovered Simpson's paradox in 1899. It was not published until 1951 by mathematician Edward Simpson, so Simpson gets the credit. A paradox is something that happens, but it should not happen. It's a real head-scratcher.

For instance, if I do a triathlon with 2016 Olympic Gold medalist Gwen Jorgenson, and I beat her in the swim, I am faster than her on the bike, I beat her in both transitions, and I beat her on the run, I should beat her in the race! If somehow she ends up beating me, that would be a paradox.

Here is Simpson's paradox in a more traditional setting, the NBA. The table summarizes the shooting percentages in the 2001-02 season in the NBA by Brent Barry and Shaquille "Shaq" O'Neal..

NBA Shooting percentages

Shot type	Brent Barry		Shaquille O'Neal	
2 point shot	237/403	58.8%	712/1228	58.0%
3 point shot	164/387	42.4%	0/1	0.0%
overall	401/790	50.8%	712/1229	57.9%

Notice Barry beat Shaq in 2-point shooting, and also beat Shaq in 3-point shots. Barry beat Shaq in both, so Barry should beat Shaq when we combine them, right? Wrong! Barry finished the season lower than Shaq on shooting percentage even though he was better in all categories.

That is the paradox. Something happened that should not happen. If you are better in the individual categories, you should be better when they're combined! But Barry wasn't. That's Simpson's paradox.

Now, why did it happen? Barry shot almost half of his shots from 3-point range, which are harder to make and therefore have a lower percentage. Shaq took only one shot from beyond the arc (and missed it), his overall average was hardly affected at all! His average stayed pretty close to his 2 point percentage of 58%, while Barry's got pulled down to below 51% because of the harder 3 point shot. It was an unequal distribution of numbers, Barry having taken many more threes than Shaq.

Here are more examples of Simpson's paradox from MLB.

Philadelphia Phillies

Player	2006		2007		Combined	
Ryan Howard	182/581	0.313	142/529	0.268	324/1110	0.292
Gregg Dobbs	10/27	0.370	88/324	0.272	98/351	0.279

Dobbs had a higher batting average than Howard in 2006 and 2007, but combined Howard was better. He was because of an unequal distribution of numbers. Howard had the majority of his at-bats in 2006 when he hit .313, and Dobbs had the vast majority of his in 2007 when he hit .272.

New York Mets

Player	2008		2009		Combined	
Angel Pagan	25/91	0.275	105/343	0.306	130/434	0.300
Carlos Beltran	172/606	0.284	100/308	0.325	272/914	0.298

Beltran was better both years, but Pagan was better overall. It was because of an unequal distribution of numbers. Beltran had the vast majority of his at-bats in 2008 when he hit .284, bringing his average down. Pagan had the vast majority of his at-bats in 2009 when he hit .306, bringing his average up and ultimately higher than Beltran.

One more baseball example. We'll look at Dustan Mohr (Twins and Giants) and Darin Erstad (Angels) during the 2003 baseball season. Here we compare their batting averages with runners in scoring position and no runners in scoring position. Take a look:

2003 Baseball Season

Player	Runners		No runners		Overall	
Dustan Mohr	19/97	0.196	68/251	0.271	87/348	0.250
Darin Erstad	9/50	0.180	56/208	0.269	65/258	0.252

Mohr hit better with runners in scoring position, and with no runners in scoring position. But, Erstad hit better overall. You can see the problem with the overall distribution of the numbers. Erstad only had 50 at-bats when he coughed up a .180 average. Since it was only 50, it did not pull his overall average down as much as Mohr's did (as Mohr had almost twice as many at-bats with runners in scoring position).

And back to the NBA for our last example. These are team (rather than individual) shooting stats from Game 5 of the 2011 NBA playoffs between the Spurs and the Grizzlies:

2011 NBA Playoffs Game 5

Team	2-point shooting		3-point shooting		Combined	
Grizzlies	38/77	49.4%	3/10	30.0%	41/87	47.1%
Spurs	32/63	50.8%	7/22	31.8%	39/85	45.9%

The Spurs had better 2-point accuracy, better 3-point accuracy, but overall were outshot by Memphis! That is the paradox. The Spurs should have been better overall.

But what have we learned from this section? Statistics is a tool in our toolbox, but not the only tool. Sometimes we cannot take the data at face value. We have to dig deeper. We have to find another tool in our toolbox.

SIMPSON'S PARADOX PRACTICE PROBLEM

In 1995,

Derek Jeter was 12 for 45. What was his batting average? _____________

David Justice was 104 for 411. What was his batting average? ___________

Who was the better hitter and why?

In 1996,

Jeter was 183 for 582. What was his batting average? _____________

Justice was 45 for 140. What was his batting average? _____________

Who was the better hitter and why?

Combining 1995 and 1996,

Jeter was 195 for 630. What was his batting average? _____________

Justice was 149 for 551. What was his batting average? _____________

Who was the better hitter and why?

One was better in 1995 and 1996 but the other was better when the years were combined. Why?

In 1997,

Jeter was 190 for 654. What was his batting average? _____________

Justice was 163 for 495. What was his batting average? _____________

Who was the better hitter and why?

Combining 1995, 1996 and 1997,

Jeter was 385 for 1284. What was his batting average? _____________

Justice was 312 for 1046. What was his batting average? _____________

Who was the better hitter and why?

One was better in 1995, 1996 and 1997 but the other was better when the years were combined. Why?

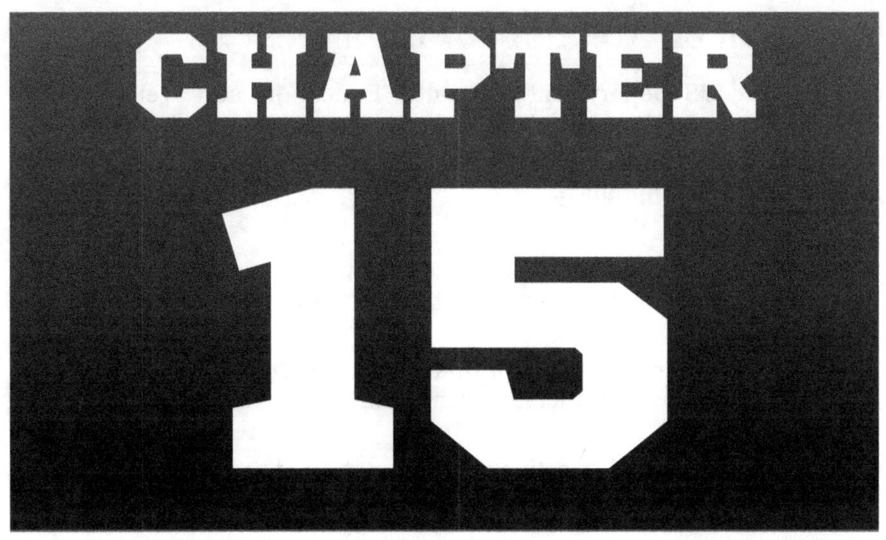

Title IX: Thou shalt not discriminate

No person in the United States shall, on the basis of sex, be excluded from participation in, be denied the benefits of, or be subjected to discrimination under any educational program or activity receiving Federal financial assistance.

— Title IX

PREVIEW: FEDERAL LAW EXPANDS SEX-DISCRIMINATION PROTECTION INTO SCHOOL SPORTS

Title IX of the Education Amendments of 1972 is a United States law enacted on June 23, 1972. The law states, "No person in the United States shall, on the basis of sex, be excluded from participation in, be denied the benefits of, or be subjected to discrimination under any education program or activity receiving Federal financial assistance." — United States Code Section 20.

Although Title IX is best known for its impact on high school and collegiate athletics, the original statute made no explicit mention of sports.

Three-prong test of compliance

In 1979, the US Department of Health, Education, and Welfare under President Jimmy Carter's administration issued a policy interpretation for Title IX, including what has become known as the three-prong test of an institution's compliance.

PRONG 1 - Providing athletic participation opportunities that are substantially proportionate to the student enrollment, OR

PRONG 2 - Demonstrate a continual expansion of athletic opportunities for the underrepresented sex, OR

PRONG 3 - Full and effective accommodation of the interest and ability of underrepresented sex.

A recipient of federal funds can demonstrate compliance with Title IX by meeting any one of the three prongs.

Springfield College has an undergraduate enrollment that is 51% male and 49% female. So, 51% of the athletic opportunities should be for men. Prong 1 of Title IX says that athletic opportunities should be proportionate amongst men and women. Women at SC should have 49% of the athletic opportunities. Do they? Does Springfield College meet Prong 1?

Prong 1 is the easiest to measure, and since a college only has to satisfy one prong of Title IX, if you can satisfy Prong 1, you are good to go!

When we do the question in class, we will use the most recent data. We will go to the Springfield College athletics website and add up the roster spots for the women's teams and the men's teams separately. That becomes our **observed frequency**. Right? We just observed the counts, or frequencies, of women and men participating in intercollegiate athletics here at Springfield College.

For the **expected frequencies**, we look at the language of Prong 1. These athletic opportunities that you just counted have to be "substantially proportionate" to the Springfield College student enrollment. For the men, since 51% of the student body is male, we would expect 51% of the athletic opportunities to be for men. We calculate 51% times the total number of roster spots for men and women combined. That is the expected frequency for the men. For the women, we enter .49 x the total number of roster spots for men and women combined.

To calculate if those numbers are substantially proportionate, we perform a **hypothesis test**. The same hypothesis test we did in Part 1 of the book. If the p-value is greater than .05, we fail to reject the null hypothesis. The null hypothesis, in this case, is that there is no difference between the observed and the expected frequencies, or if they are different, it is due to random variation.

In the context of this problem, if the p-value is greater than .05, we would say that the differences between the observed values and those we expected to see are due to random variation. We would not have enough evidence to say the Springfield College is out of whack, and therefore the College's athletic opportunities are substantially proportionate to the undergraduate student body.

If our p-value is less than .05, we would reject the null hypothesis. We would conclude we have enough evidence to say the Springfield College is violating Prong 1 of Title IX, that the College is not providing enough athletic opportunities for the underrepresented sex.

If that was the case, it does not mean that Springfield College is in violation of Title IX. Any college or university can satisfy Title IX through compliance with just one of the three prongs. They do not have a legal requirement to satisfy all three.

Does Springfield College comply with Title IX?

Does Springfield College comply with Title IX?

Enter data in Minitab

Label C1 Observed Counts SC Athletes. Label C2 Sex and C3 will be Undergraduate Proportions.

The observed counts are the roster spots that we counted up for men and women. Sex is the categorical variable, and we can enter Male and Female. The undergraduate proportion is tricky. First Google "percent undergrad female Springfield College". The first link will tell you that we are 49% female (and therefore 51% male). BUT, we want percent in decimal form. To convert percent to a decimal, move the decimal point two places to the left. 49% becomes .49. And .49 is what we enter in Minitab (and .51 for the men).

Choose Minitab options

Choose Stat from the menu bar, Tables, Chi-Square Goodness-of-Fit Test (One Variable). Then enter the observed counts in C1 and the category names in C2. The test defaults to equal proportions, which was fine when we did the first exam. But now we have unequal proportions. (In SC's case, 51% vs. 49%, notice they are unequal.) Click on specific proportions and put C3 in that box.

Note: Be careful. On the first exam, we did the exact same chi-square goodness-of-fit test, but we used the default, Equal proportions. When we do chi-square for Exam 2, it is hard to remember to switch to Specific proportions because Minitab defaults to equal, AND we used the default the first time through. Go slow when you do the test on Exam 2, the only change is to use Specific proportions instead.

We will do Springfield College in class, so we have the latest data. But Boston University is next. I will show you Minitab as we work through BU. You will then have the answers for BU. Then you have Providence College, Vermont, and Hartford for practice.

GUIDED EXAMPLE FOR TITLE IX WITH BU

Boston University was a national power in football at the I-AA level as it was called then, even playing in the national championship game. Then BU dropped football as part of their Title IX efforts to support women's teams. Does BU now satisfy Prong 1 of Title IX?

I went to the BU website and found all the teams for varsity athletics for women and men, and I added up all the roster spots.

BU Sports 2014

Men	Sport	Women
14	Hoops	10
19	XC	22
0	FH	18
0	Golf	10
26	Hockey	25
50	LAX	30
44	Rowing	75
35	Soccer	27
0	SB	21
26	Swim/Dive	29
10	Tennis	9
46	T&F	56
270	Total	332

The athletic department looks good so far. They have 62 more roster spots for women. But it turns out the undergraduate enrollment at BU is 60% women and only 40% men. They should have more opportunities for women. Women should have about 60% of the opportunities. If the women do have about 60%, then BU satisfies Prong 1 of Title IX. To test that they do, we perform a chi-square goodness-of-fit test, which we have done in Part 1 of the book.

1.   **H_0:** For sex, the proportion of undergrads = proportion of athletes
2.   **H_A:** For sex, the proportion of undergrads ≠ proportion of athletes

Enter data in Minitab

Now we enter our data into Minitab. In C1 we will put the observed counts. In C2 the sex. In C3 the proportions. With the proportions, Minitab will not accept the percents. We have to enter the percent as decimals. Just move your decimal point two places to the left to make it into a decimal number.

   Important note: Remember it defaults to equal proportions. Click on specific proportions to change this.

Choose Minitab options

Choose Stat in the menu bar then choose Tables then Chi-square Goodness of Fit. For Observed counts double click on C1. For Category Names double click on C2. For **Specific proportions**, double click on C3. Then OK.

Done!

The result of the test is here:

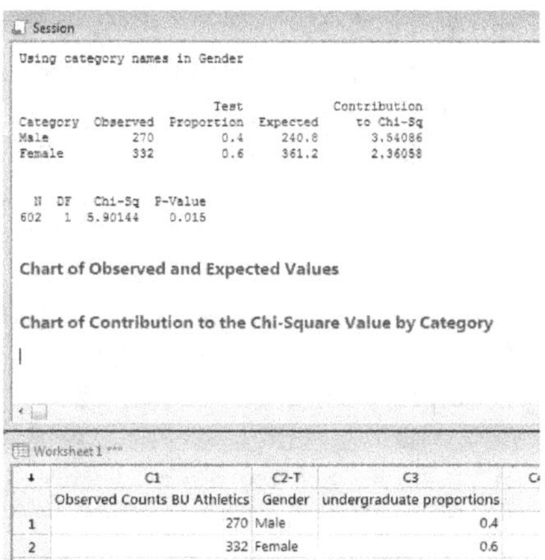

	C1	C2-T	C3	C
	Observed Counts BU Athletics	Gender	undergraduate proportions	
1	270	Male	0.4	
2	332	Female	0.6	

3.   **Test statistic and its value:** $\chi^2 = 5.90$
4.   **p-value:** .015
5.   **Your statistical conclusion:** Since the p-value < .05, I reject H_0

6. **Your interpretation:** The differences we see in BU's numbers are not due to random variation. BU does NOT satisfy Prong 1 of Title IX.

    Minitab output note: We counted 270 roster spots for the men. In the fourth column, Minitab calculated that the men should have had 240.8 roster spots. This was our first clue BU was in trouble. They allocated 29.2 more roster spots to the men than they deserved.

The women were our second clue. They were allocated 332 roster spots but should have had 361.2. The women were short 29.2 roster spots.

PRACTICE PROBLEMS

We will the 6-step hypothesis test answer format for each question.

Hypothesis Test
1. **H$_0$:**
2. **H$_A$:**
3. **Test statistic and its value:**
4. ***p*-value:**
5. **Statistical conclusion:**
6. **Interpretation:**

Providence College has no football program, which means the College has only 204 varsity roster spots for men. The women have 179 roster spots but makeup 56.5% of the undergraduate population, while the men make up the balance (43.5%).

Perform a hypothesis test to determine if Providence College satisfies Prong 1 of Title IX.

Does the University of Vermont (UVM) satisfy Prong 1 of Title IX? They dropped football and now have only seven sports for men, which they needed to do as their undergraduate population is only 43.7% male and, therefore, 56.3% female. The observed frequency of athletes at UVM is 185 for the men and 216 for the women.

Perform a hypothesis test to determine if UVM satisfies Prong 2 of Title IX.

3 Does the University of Hartford satisfy Prong 1 of Title IX? 51% of their undergraduates are female, and 49% are male. But, even without football and wrestling (two sports they have dropped), they have 146 female athletes and 168 male athletes.

Perform a hypothesis test to determine if the University of Hartford satisfies Prong 1 of Title IX.

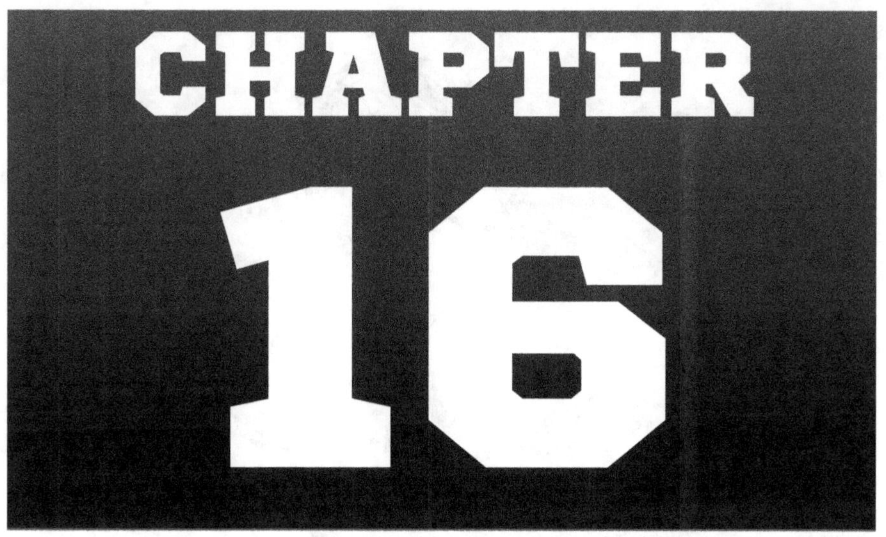

Trending now in sports

You miss 100% of the shots you don't take.

— *Wayne Gretzky, NHL Hall of Famer*

PREVIEW: A GRAPH TO INVESTIGATE DATA OVER TIME

A **time-series plot** graphs data that was collected over time, thus the name time-series plot. For example, a rookie running back in the NFL may start his rookie season with huge numbers. But after four exhibition games, then a sixteen-game regular season with only one bye week, then potential playoff games against bigger, faster, and stronger players than he faced in college, the extra strain of NFL competition will wear him down. If we plotted his yards per game for his rookie year over his 16 regular-season games, we would probably see a decline in the number of yards per game. We could do the same with his yards per carry.

This plot of time-series data will reveal trends to opposing teams and his own coaches. These are trends over time. The x-axis is always time, whether it be years or weeks or days or hours. The y-axis is always the data, in our example yards per carry or yards per game. You will put a dot on the graph representing each data point and then, at the end, connect the dots with a line to see the trend.

For example, let's examine Wayne Gretzky's incredible career. He scored 894 goals in 1487 games, better than one goal every two games for 21 years!

Enter data in Minitab

Enter your numbers to analyze into C1. In this case, I copied Gretzky's career goals per season and pasted them into C1.

Choose Minitab options

Then go to "Graph", "Time Series", and keep it "Simple". Double click on C1 to move it into the series box, then click labels. For labels, I added titles and subtitles to put the numbers into context, subtitle 2 became my name, and in the footnote, enter your interpretation of the graph. Namely, what do you see happening throughout Gretzky's career?

Then click OK.

Done!

For now, this line just gives us the trend. It tells us whether our data is rising or falling, or even flat. I listed my interpretation of Gretzky's career scoring in the footnote of his graph below. It says, "After a slow start, Gretzky peaked in year three, then has been trending down over time."

When you do a time-series graph, look for the overall trend, not the year-to-year ups and downs

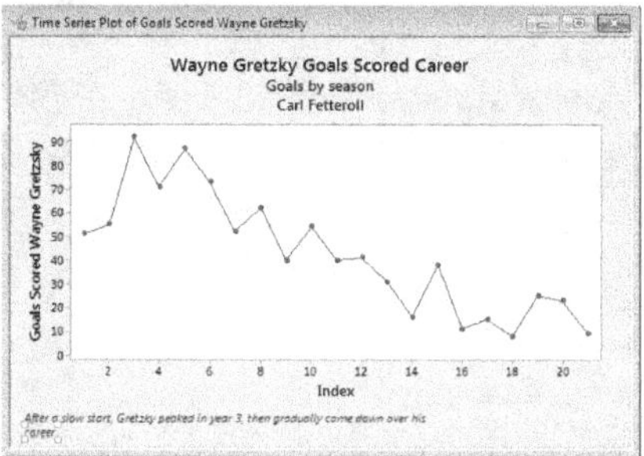

Another use for a time series plot is a comparison. For instance, David Ortiz is retired. You can compare any two players you want. There are two free agents with a history of power-play for the Blue Jays: Edwin Elpidio Encarnación and Jose Bautista. Encarnación has 310 home runs in 12 seasons, while Bautista has 308 home runs in 13 seasons. While their careers are similar, we can use a time series plot to look at the recent trends, in this case, for power.

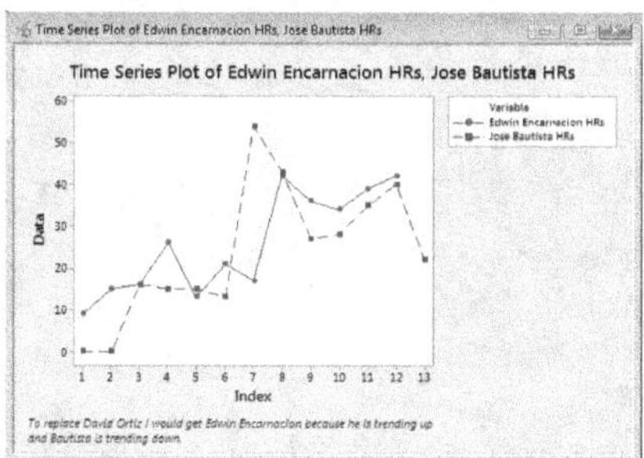

Encarnación is a better choice to sign as a free agent because he is trending up the last two seasons while Bautista had a large drop-off this year.

PRACTICE PROBLEM

1 Sidney Crosby is one of the premier players in the NHL and was the team Canada captain for the Sochi Olympics. Look up Sidney Crosby's stats. Create a time-series plot of his goals scored throughout his career. Label the graph. Then use the footnote of the graph for your interpretation.

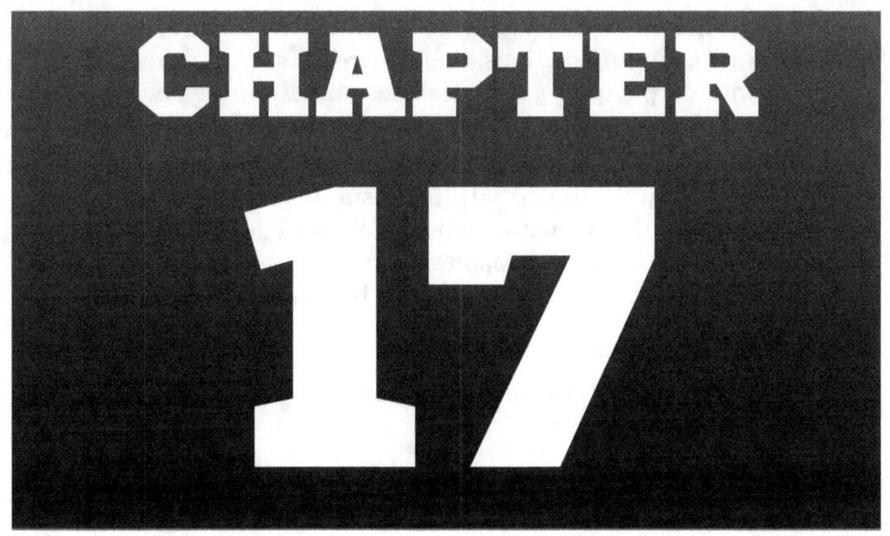

Gaze into your crystal ball

It's tough to make predictions, especially about the future.

— *Yogi Berra, American icon*

PREVIEW: PREDICT THE FUTURE WITH REGRESSION

Making predictions about future performance

After you graduate from Springfield College with your minor in Sports Analytics, you may get a job as general manager of the Boston Red Sox. Your first task is to upgrade the team's offense. You are offered Edwin Encarnación, and you wonder how he would fit in with your line-up in Boston. You can plug his stats into your team's statistical model, and you predict Encarnación's performance for the next season with the Boston Red Sox. That prediction you made? That is **regression**. Simple, right? Now let's see how regression works.

To start with, we need two quantitative variables, like ERA and strikeouts, or OBP and OPS. Go to baseball-reference.com and find the 2015 stats for the Boston Red Sox.

I mentioned OBP and OPS above, so let's use those. OBP is On Base Percentage, and OPS is On Base Percentage plus Slugging. First, I want to see if I have a **linear relationship** between those two variables before I use them for making predictions. A linear relationship means it can be described with or pictured as a line. The points form a line when one value goes up or down. The other goes up or down. They're tied together. If the relationship is linear, we can use one variable to predict the other.

There are two ways to verify if there is a linear relationship. One is with a graph. The other is mathematically. We will do both. And just like categorical and quantitative variables, we will start with the graph first. But this time the graph is a new one for us. It is called a **scatterplot**.

Enter data in Minitab

To graph a scatterplot I copied the columns labeled OBP and OPS into Minitab (see below):

```
--------------  8/15/2016 10:27:08 AM  --------------

Welcome to Minitab, press F1 for help.

Scatterplot of Red Sox 2015 OPS vs Red Sox 2015 OBP


Correlation: Red Sox 2015 OBP, Red Sox 2015 OPS

Pearson correlation of Red Sox 2015 OBP and Red Sox 2015 OPS = 0.955
P-Value = 0.000


Correlation: Red Sox 2015 OBP, Red Sox 2015 OPS

Pearson correlation of Red Sox 2015 OBP and Red Sox 2015 OPS = 0.820
P-Value = 0.000
```

	C1	C2	C3
	Red Sox 2015 OBP	Red Sox 2015 OPS	
1	0.319	0.712	
2	0.307	0.693	
3	0.356	0.797	
4	0.355	0.776	
5	0.292	0.658	
6	0.291	0.717	
7	0.341	0.820	
8	0.288	0.647	
9	0.360	0.913	
10	0.349	0.727	
11	0.335	0.832	
12	0.327	0.813	
13	0.337	0.664	
14	0.347	0.831	
15	0.238	0.439	

Choose Minitab options

Choose Stat from the menu bar. Then Regression and click on Fitted line plot. The fitted line plot is the scatterplot. On the x-axis you will put the **independent variable**, also known as the explanatory variable, and on the y-axis you will put the **dependent variable**, or the response variable. Minitab asks for y first, which seems backwards, but just make sure you click the y first because of that.

Which variable is y? It is the result. Like comparing 5K time and miles per week. Your 5K race result *depends* on how many training miles you ran. The 5K time is the y. Or hours studying and your grade on the next exam. Your grade is the result and will depend on how many hours you study. Your grade is the y. How about OBP and OPS? OPS is a higher-level variable, so OPS depends on your OPB. OPS will be our y and OBP will be our x.

Note: There are times when one variable is not the result of the other, like comparing ERA and WHIP. You could easily argue that ERA depends on WHIP, but the same argument holds true for WHIP depending on ERA. In that case, it does not matter which is y or x. In class, we will put the first variable in as x and the second as y so we can all calculate the regression equation the same, but you would not be wrong in doing it the other way.

Double click on OPS to move it to y and OBP to x. Click OK.

Done!

The scatterplot will look like this:

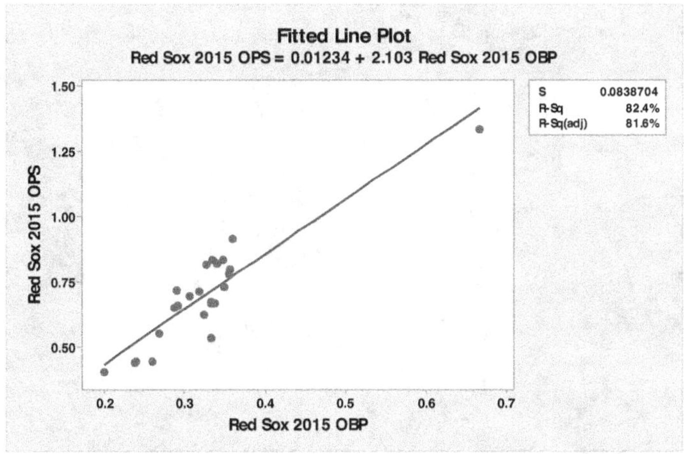

Remember, we are looking for a relationship between OBP and OPS. If we see a pattern in the dots on our scatterplot, there is a relationship.

In the graph above, we can see the dots clustering. We do see a pattern. As Minitab shows, we could put a line through the data points. We have a linear relationship! Yes, a relationship does exist between OBP and OPS.

The closer the dots are together, the stronger the relationship. If the dots formed a line, it would be a very strong relationship. Our dots are close but not that close. We would say we have a strong, linear relationship.

One more aspect of this new relationship we have formed is direction. If the line is going up (from left to right) as above, it is **positive**. If the line is going down from left to right, it is **negative**. We have a strong, positive, linear relationship.

Note: Positive is not good, and negative is not bad. Positive in graphing means that as one value goes up, the other value goes up. Or as one value goes down, the other value goes down. They go in the same direction. That makes sense. In baseball, when a player gets hot and his

OBP goes up, we would expect his OPS to go up as well. Also, when he slumps and his OPB goes down, we would expect his OPS to go down, too. We would have guessed before we started that, if a relationship existed, it would have been positive.

An example of a negative relationship would be as one value goes **up**, the other goes **down.** They go in different directions. Like training miles per week and 5K time. If you can squeeze in more training, the miles per week increases. Increasing training miles hopefully leads to *decreased* 5K times because you are running faster! As one goes up, the other goes down. We call that a negative relationship. Not in a bad (or negative) way though.

Remember, our ultimate goal was to predict a player's OPS when he is plugged into the Red Sox lineup, using his OBP. We can make that prediction IF we have a linear relationship. Based on the graph we generated on Minitab, we decided we do have a relationship, specifically a strong, positive, linear relationship.

But, if the dots were further apart we might not be certain of a linear relationship. We need a mathematical way to check, a method that is black and white and leaves us with no doubt about whether a relationship exists. That tool is called **correlation**. Look at the root of the word. Cor- means "together" + relation. It it is a measure of a relationship between two variables.

To calculate correlation on Minitab we have two options. On the graph Minitab created, it gave us r^2 (R-Sq in the upper right). In statistics it is standard to use r for correlation or, more specifically, the Pearson correlation coefficient. We could just take the square root of that value and that would give us r. Or on Minitab we can choose Stat, Basic statistics, Correlation, and put the two variables in the box, and say OK. It will give us the **Pearson correlation value**. (Karl Pearson was a big deal in mathematical statistics. He developed the hypothesis testing we do in this course. The official name for the chi-square test is the Pearson chi-square test. He is also generally credited for the histogram.)

Here is the Minitab output:

Correlation OBP, OPS

```
Pearson correlation of OBP and OPS = 0.908
```

But what does this number mean? r can take on any value between -1 and 1, inclusive. The closer r is to -1 or 1, the stronger the relationship. The closer r is to 0, the weaker the relationship. In our example, we have an r of .908, which is pretty close to 1, we have a strong relationship. Also, r is positive, we have a strong, positive relationship. Which agrees with our graph.

Basically r measures the strength and direction of a linear relationship. We can only use r when the relationship is linear. And r will match the slope. If r is positive, the slope is positive, and the regression line goes up from left to right. If r is negative, the slope is negative, and the regression line goes down from left to right.

Which is stronger, an r of .89 or an r of .65? .89 is because it is closer to 1. How about an r of -.77 or an r of -.92? -.92 because it is closer to -1? The closer you are to -1 or 1, the stronger the relationship. How about -.71 or .71? Neither! They are equally strong because they are the same distance from zero.

Now r^2. r^2 is another measure of the strength of our model. When you loaded the data into Minitab for OBP and OPS, and then ran regression, you created a model. That model is what we will use to predict OPS from OBP now that we have determined through the graph, and confirmed with r, that a linear relationship exists.

In this case r² gives us the percent of the variation in OPS that can be explained by the batter's OPB. In general, it is always the percent of the variation in y that can be explained by x. Those y and x values will change with each problem we do.

We can find r² in two ways. First the easy way, it is on your scatterplot! Of the two choices, use R-Sq, not R-Sq(adj), as R-Sq represents the data better. Second, R² is literally r squared (r²). Yes, you can do r times r on your phone.

On your graph you also have at the top the actual regression equation. In our case, it is:

Red Sox 2015 OPS = .01234 + 2.103 Red Sox 2015 OBP

What does this equation mean? Remember, we are trying to predict OPS for a new Red Sox player. Well, that is the left side! "Red Sox 2015 OPS =." That is where our prediction will show up. The next is .01234. That is the y intercept, which we have to be very careful with in regression, because it is either outside of our data's range or just plain meaningless. For instance, if a player has an OBP of 0, the y intercept says he will have a predicted OPS of .01234! Impossible. Look at any player on the Red Sox, and you will see plenty of pitchers with an OBP and slugging of 0, when you add them together, 0 + 0 = 0, right? Not .01234! Because of this, we rarely use the y intercept with regression, and if we do, we do it with caution. We won't see it again in this course.

But the next number we will see again. 2.103 is the slope. Notice the slope is positive, as it should be since the regression line goes up from left to right. And the positive slope agrees with r, in terms of both being positive. But the slope also has a deeper meaning, it tells us how much our y will change based on a unit change in x. Huh? Well, in our case, if a player's OBP goes up by 1, his OPS goes up by 2.103! A hitting coach who improves a player's OPB, will affect his OPS by more than twice the improvement for OBP.

Now we can try to predict a player's OPS for an OBP of .360. On **Minitab** you will choose Stat, Regression, Regression, Fit the regression model. Then OPS will be your response, and OBP will be your predictor. Then say OK. The next time you do Stat, Regression, Regression, the Predict option will be available to you. Use that and type in .360, then OK. Your model will give you your prediction for a player's OPS, using any OBP numbers you type in! On Minitab it is labelled as "fit" in the output.

Based on the "Fit" number below, we would predict a new Red Sox who brings an OPB of .360 to the term will contribute an OPS of .769 to the Sox offense.

Regression Analysis: OPS versus OBP

Prediction for OPS

```
Regression Equation

OPS = .0123 + 2.103 OBP

Variable    Setting
OBP           .36


Fit         SE Fit        95% CI                95% PI
.76949      .018596   (.73092, 080805)    (.59133, 094765)
```

By hand you will plug .360 into your regression equation. Simply multiply .360 times 2.103. Take that answer and add .01234. You get .76942. They agree!

Now, are you comfortable making that prediction? Look at your Red Sox data for 2015. .360 is in the range of your data, at the very top, but in the range. Yes, you are comfortable. If I ask you to make a prediction for an OBP of .370, you could, but you would be uncomfortable. Our model is based on OBP's up to .360, we do not know how the model would react to data points that high. As long a I give you values within the range of your data, you are comfortable. Outside of that, you are not comfortable. And specifically, the x values, the input.

When you look at the Red Sox data, one of the players had an OBP of .360, then why am I asking you to predict that? We already know his OPS. First, I could have asked you to predict OPS for an OBP of .359, and no Red Sox had an OBP of .359. Second, now I can measure how well the .360 player did. Our model says he should have an OPS of .769, he actually had an OPS of .913, much higher! He outperformed our model, meaning he over performed in 2015.

The difference between his actual performance (.913) and his predicted (.769) is called his **residual**. He had a positive residual of .144, and anytime a player has a positive residual they over-perform, and a negative residual means they underperform.

Note: Look on the scatterplot earlier, and notice the regression line goes through the data points. The difference between any data point and the line is the residual. Can you guess which dot belongs to our .360 hitter based on that? It was not arbitrary how Minitab placed that line. Minitab calculated the residuals for each possible line, squared the residuals, and added them up. The line that had the smallest total of squared residuals was rewarded with showing up on our graph and the honor of being called the regression line.

GUIDED EXAMPLE FOR REGRESSION WITH THE BOSTON BRUINS

Determine if there is a relationship between assists and points for the 2019-20 Boston Bruins.

To answer the question, we will follow the 15 steps below.

15-STEP ANSWER FORMAT FOR REGRESSION

Graph

1. **Create a fitted line plot.**
2. **Describe the relationship.** (On the exam, put this in the footnote and print out the graph. (Don't forget your name!))

Correlation

3. **Calculate the correlation r.**
4. **Interpret r.**
5. **Calculate r^2.**
6. **Interpret r^2.**

Regression

7. **What is the regression equation?**
8. **Make a first prediction.** Ondrej Kase played his four-year career with Anaheim. The Bruins picked him up at the trade deadline in 2020 to fill a need as a second-line winger. How many points would

you predict for Ondrej in his first full season? He averaged 13.25 assists per season with Anaheim.

9. **Are you comfortable with this prediction and why?**

10. **Make a second prediction**. Leon Draisaitl of Edmonton led the NHL in assists in 2019-20 with 67. And, like Kase, another 24-year-old. IF the Bruins had been able to sign Draisaitl instead of Kase, how many points would you predict for Draisaitl in the Bruins lineup?

11. **Are you comfortable with this prediction and why?**

12. **Make a third prediction**. Blake Wheeler used to be a right-winger with the Bruins in his younger days. Now he is a productive player for the Winnipeg Jets. In 2019-20, he had 43 assists. If he had those 43 assists with the Bruins, how many points would you predict for him?

13. **Are you comfortable with this prediction and why?**

14. **What is the residual?**

15. **Interpret the residual**.

Graph

1.Create a fitted line plot

We know there are two ways to determine if a relationship exists: a graph and the statistical concept of correlation. We will do the graph first. The graph is a scatterplot. We find it in Minitab under regression. Notice 8, 10, and 12 above all ask for predictions. And regression is a fancy name for predictions!

Enter data in MInitab

First I searched "2019-20 Bruins stats". From the nhl.com>bruins>stats link I copied the Bruins assists into C1 and the Bruins points into C2.

Choose Minitab options

Then I chose Stat from the menu bar. I click Regression on the drop down menu. The first option is Fitted line plot, just what Step 1 is asking for.

I click Fitted line plot. I enter the higher-level variable (the one furthest to the right in the data) in Y. Points is furthest to the right, so double-clicking Points moves the data over. That leaves Assists for x.

Note: in a fitted-line plot, the y is always the dependent variable, y *depends* on x. x is the *independent* variable. We will also call y the response variable when we make predictions and x the predictor variable. Then hit OK.

Done!
Your Minitab output will look like the image on the next page. That's our answer for Step 1.

For Step 1 on the exam, you will print the graph, but first you will enter the answer to Step 2 in a footnote before printing.

2.   **Describe the relationship.**
The description involves form, direction, and strength. The form for the Bruins is linear. Why linear? Because we can put a line through the dots.

Direction looks at the line. If the line is going up (like ours is), it is positive. Our direction is therefore positive. If the line had been going down it would be negative.

The strength is strong. The closer the dots are to the line, the stronger the relationship. The more scattered the dots are, the weaker.

For Step 2 on the exam, in the footnote of the graph, you will write *"The relationship is linear, positive and strong."* If you had said "very strong" that is fine! You are right. Qualifiers are allowed. *And in the subtitles, you will put your name.*

The Bruins have a linear, positive and strong relationship between assists and points. That means the more assists a player has, the more points he will accumulate. And the relationship is very strong. We can see that from the graph.

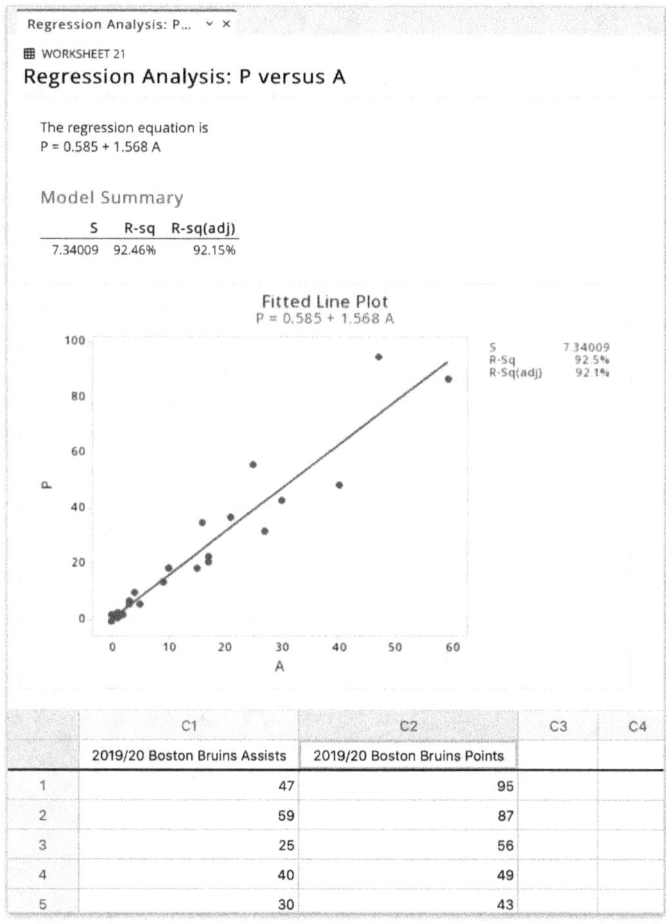

	C1	C2	C3	C4
	2019/20 Boston Bruins Assists	2019/20 Boston Bruins Points		
1	47	95		
2	59	87		
3	25	56		
4	40	49		
5	30	43		

Why do we need a statistical number to determine if there is a relationship? The graph was pretty easy, right? Well, what if the dots were very scattered? How do you tell when a relationship goes from strong to regular to weak to non-existent? We will all interpret graphs differently, it is the nature of the business.

But if a player has a salary of $1,000,000, that number means the same to all of us. Step 3 of the problem is about finding that number that we can all interpret the same way. And that number is ...

Correlation

3. Calculate the correlation r = .962
The Pearson Correlation Coefficient. Pearson is the guy who popularized it. Correlation means the relationship between two variables. And coefficient means number. In our Minitab output, it

will be called r.

Minitab calculates r when you choose Stat, Basic statistics, Correlation …. Put both assists and points into the box (order does not matter) and say OK. Minitab tells us r for the Bruins is .962.

For Step 3 on the exam we would enter $r = .962$.

4. **Interpret r:**
    Now we have to interpret r. It is the strength and direction of a linear relationship. For r, it can range from $-1 < r < 1$, and the closer we are to -1 or 1 the stronger the relationship. The closer to 0, the weaker the relationship. Our r is pretty close to positive 1, so a strong, positive relationship between assists and points for the 2019-20 Bruins.

    For Step 4 on the exam we write: *"There is a strong, positive relationship between assists and points for the 2019-20 Bruins."*

5. **Calculate r².**
    Now we can calculate r^2. It is easy to calculate because it is on the Minitab print out AND the graph above. Never use the adjusted value (R-sq adj), just use the r-sq value. From the worksheet, r-sq = 92.46. From the graph, r-sq is 92.5. Same value, just rounded on the graph. And these values are percents.

    For Step 5 on the exam we would enter $r^2 = 92.5\%$.

6. **Interpret r²:**
    To interpret that value, look at our graph. In general, r-sq is the percent of the variation in y that can be explained by x. Our y is Points. Do all the Bruins have the same amount of points? No. There is variation within the team. They all have different point totals.

    Why do the players all have different point totals? Our graph says it is because of their assists. The more assists a player has, the more points they will have. Now we can answer Step 6 specifically for the Bruins that 92.5% of the variation in points can be explained by the players' assists.

    For Step 6 on the exam we would write *"92.5% of the variation in points can be explained by the players' assists."*

Regression

7. **What is the regression equation?**
    You put the Bruins assists in C1 and the points in C2 and then hit

Regression in Minitab. You created a model we can use for predicting points. The model is represented by the regression equation. And the regression equation is on your worksheet AND your graph. It looks like this:

P = .585 + 1.568A where P stands for Bruins point totals and A stands for assist totals. And this regression equation looks a lot like our old linear equations in Algebra. This means the number attached to A is ... yes! The slope! What the slope means is for every extra assist a Bruins player has, he will end up with 1.6 more points. And the 1.568 is positive, and that sign will match the sign of our r. Notice that r was a positive .968, and the slope is positive as well. They will always agree.

For Step 7 on the exam we would write: *P = .585 + 1.568A*

Now we can use the regression equation to make predictions.

8. **Make a first prediction.**
   Ondrej Kase played his entire four-year career with Anaheim. The Bruins picked him up at the trade deadline in 2020 to fill a need as a second-line winger. How many points would you predict for Ondrej in his first full season? He averaged 13.25 assists per season with Anaheim.

You have your prediction equation, just plug in his 13.25 assists and solve for P, like this:

Regression = .585 + 1.568A
Plug in: = .585 + (1.568 x **13.25**)
Multiply: = .585 + **20.776**
Add: = **21.361**
Round: = **21** (leaving it as 21.361 is fine too!)

We predict Kase will score 21 points with the Bruins in 2020-21.

You are thinking, OK, that math is not bad, but why can't Minitab calculate the prediction for us? You are right! Minitab can. In Minitab choose Stat, Regression, Regression and you will notice in the drop-down menu in Minitab is Prediction. BUT, it is grayed out. To activate it, we have to do: Stat, Regression, Regression, Fit the model first.

When you fit the model, do the same x and y as you did with the graph. Points (P) are the y and assists (A) are the x. Then hit OK and Minitab has loaded the model and prediction will be available. Choose Stat, Regression, Regression, Prediction, and load in Kase's 13.25 assists. Then hit OK. On your worksheet near the bottom will be the fit number. That is your prediction. In this case, it will be the same 21.361 you did by hand. Feel free to use either method on the exam!

9. **Are you comfortable with this prediction and why?**
Comfortable means are you comfortable using your model. It depends on your input. If the value is in the range of our input, we are comfortable. If it is outside of that range, we are not comfortable.

Our input. Our input is the x. Our x is Assists. For the Bruins, the assists range from 0-59. I am comfortable making any prediction that is in that 0-59 range (including 0 or 59). Kase had 13.25. He is between 0 and 59. I am comfortable with this prediction because Kase is in the range of the data.

10. **Make a second prediction.**
Leon Draisaitl of Edmonton led the NHL in assists in 2019-20 with 67. And, like Kase, another 24-year-old. If the Bruins had been able to sign Draisaitl instead of Kase, how many points would you predict for Draisaitl in the Bruins lineup?

You can use the same formula and your phone, and just plug in 67 for A instead. Or in Minitab, choose Stat, Regression, Regression, Prediction and change the 13.25 to 67. Then hit OK. You will get a fit on your screen of 105.641, or 106 if you want to round. We predict he would score 106 points with us.

11. **Are you comfortable with this prediction and why?**
This time I am not comfortable. Draisaitl's 67 is out of the range of our inputs. Remember, our inputs ranged from 0 to 59 (Brad Marchand by the way), anything over 59 is too high. Making the 67 too high. I am therefore NOT comfortable predicting his points.

12. **Make a third prediction.**
Blake Wheeler used to be a right-winger with the Bruins in his younger days. Now he is a productive player for the Winnipeg Jets. In 2019-20, he had 43 assists. If he had those 43 assists with the Bruins, how many points would you predict for him?

Choose Stat, Regression, Regression, Predict, and enter his 43 assists.

Hit OK and Minitab finds his predicted points (fit) as 68.009, or 68 points.

13. Are you comfortable with this prediction and why?
Yes, because 43 (the input) is between 0 and 59, I can make the prediction and feel comfortable doing it.

My favorite player is Patrice Bergeron. He had 25 assists in 2019-20. How many points would you predict for Patrice? Plugging in 25 into Minitab spits out 39.785, or about 40 points.

But, you think since Bergeron was on the team last year, we know his point total, we do not need to predict it. But our last question is what is his residual? The residual is the difference on the graph between the point and the line. Mathematically we find his residual by subtracting his actual minus his predicted. His actual points were 56, and our prediction above was 40. 56 − 40 = 16, Bergeron's residual.

14. What is his residual?
Ahhhh, residual. We know it is actual minus predicted. But what about Minitab? Can Minitab do residual? Yes!

Choose Stat, Regression, Regression, Fit the plot, Options, and check off Residual. Hit OK and Minitab will automatically fill in C3 with everyone's residual. Next to Bergeron, you will see a 16 in C3! Now we need to interpret his residual. A positive residual means he is above the line. Above the line means he over-performed. A negative residual is below the line. A negative residual means h underperformed.

15. Interpret his residual.
Since the residual is positive, Bergeron over-performed in 2019-20.

One more while we are here, look at Torey Krug. What is his residual? We look at C3, and next to Torey we see a -14.3. Torey scored 14.3 fewer points than the model predicted. Because it is negative, Krug underperformed.

And that ends regression!

PRACTICE PROBLEMS

1 The Celtics are 9-2, the best record in the entire NBA. The number one scorer is Kyrie Irving, and he is only 18th in the league. We will use NBA scoring to answer this question.

Google "NBA stats", then click on the ESPN link. Under Offensive Leaders (the first box), click Complete leaders. Up top, change season to 2016-17, change League to East, change Position to Point guard. You should have 26 players. Copy two columns into Minitab for all 26 players: MPG (minutes per game) and PTS (points).

We will investigate to determine if there is a relationship between minutes and points for NBA point guards.

Graph

1. **Create a fitted line plot**.
2. **Describe the relationship.** If it exists. Put this in the footnote and print out the graph. (Don't forget your name!))

Correlation

3. **Calculate the correlation r**.
4. **Interpret r**.
5. **Calculate r²**.
6. **Interpret r²**.

Regression

7. **What is the regression equation**?
8. **Make a first prediction.** Shane Larkin has been a contributor for the Celtics this year, playing 10.1 minutes per game (MPG). If we had Larkin last year, what would you have predicted his points would have been?
9. **Are you comfortable with this prediction and why?**
10. **Make a second prediction**. In 2016, Isaiah Thomas played 33.8 minutes per game and led the East in scoring. How many points would you predict for him based on the model you created?
11. **Are you comfortable with this prediction and why?**
12. **Make a third prediction**. John Wall in 2016 played 36.4 minutes per game (MPG). What would you predict for his points?
13. **Are you comfortable with this prediction and why?**
14. **What is his residual?**
15. **Interpret his residual.**

Switching to volleyball, the men's team has been ranked number 1 in the country since the preseason. The team is led by another freshman phenom, Eli Gabriel Irizarry Pares, who has totaled an astounding 378 assists, and the season is not over yet. So far he has led the team to a 23-1 record, losing only 6 sets all year!

Go to Springfield College Men's Volleyball, click on More, then Statistics, then Lineup. Copy the two columns (K and PTS) into Minitab.

We will investigate to determine if there is a relationship between kills (K) and points (PTS) for our Springfield College men's volleyball team.

Graph

1. **Create a fitted line plot**.
2. **Describe the relationship.** If it exists. Put this in the footnote and print out the graph. (Don't forget your name!))

Correlation

3. **Calculate the correlation r**.
4. **Interpret r**.
5. **Calculate r²**.
6. **Interpret r²**.

Regression

7. **What is the regression equation?**
8. **Make a first prediction.** Matthew Knigge of Vassar is second in the country with 336 kills. If we had Matthew on our team, how many points would you predict for him?
9. **Are you comfortable with this prediction and why?**
10. **Make a second prediction.** Last year Trevor Mattson had 142 kills as he ended his career at Springfield College. If he had played this year, how many points would you have predicted for him?
11. **Are you comfortable with this prediction and why?**
12. **Make a third prediction.** Eli Gabriel Irizarry Pares has 44 kills this year, how many points would you predict for him?
13. **Are you comfortable with this prediction and why?**
14. **What is his residual?**
15. **Interpret his residual.**

Which is the best predictor?

You only get a once in a lifetime opportunity so many times.

— Ike Taylor, Steelers cornerback

PREVIEW: FIND THE BEST WAY TO PREDICT THE FUTURE

In the Chapter 17 we learned about residuals. The residual is the difference between the actual values and the predicted values. The predicted values came from the regression model (the line) Minitab drew for us. On the graph we could see residuals. We could see how far away each point was from the regression line. Some points were above the line meaning their residuals were positive. Some below and had negative residuals. And some points were on the line. They had a residual of 0.

Remember how I said a test statistic turns an entire set of data into a single number? We can do the same things with residuals. We can gather all those residuals together into a single number. That is the **root mean square error** (RMSE). RMSE (also called the root mean square deviation, RMSD) is the average of all the residuals. It measures the difference — the error or deviation — of the observed values from the predicted values (the values represented by the line).

To do it by hand you would subtract each actual value from the predicted value. Square them to get rid of the negatives. Add them all up. Divide by the number of values which gives the average (mean) of how far the data deviates from predicted. Then take the square root. Here's what the formula looks like:

$$\text{RMSE} = \sqrt{\frac{\sum_{i=1}^{n} (\text{Actual}_i - \text{Predicted}_i)^2}{n}}$$

Where Actual is the real-life values that *actually* occurred and Predicted is the values our model *predicted* would happen.

Fortunately Minitab will calculate RMSE for us.

When we did the scatterplot in regression (Chapter 17), we used the fitted line plot. There was one value on the graph summary we did not use in regression, and that was S. Well, S is short for RMSE! When we run regression and generate our graph, Minitab calculates RMSE for us, and then displays the result as S.

Let's try an example. When I look at the stats for MLB hitters in the American League in 2016, I notice the final column is WAR (wins above replacement). It is a relatively new statistic that's supposed to sum up a player's total contributions to their team. WAR is the number of additional wins a player contributes above (or below if negative) what's expected from a replacement-level player.

I am curious which of the old stats is the better predictor of WAR. I loaded Average (AVG), OBP, Slugging (SLG), OPS and WAR into Minitab (C1 through C5 on the MInitab output to the left).

Worksheet 1 ***

↓	C1	C2	C3	C4	C5
	AVG	OBP	SLG	OPS	WAR
2	0.325	0.388	0.457	0.845	5.2
3	0.323	0.439	0.578	1.018	9.2
4	0.320	0.406	0.629	1.035	4.2
5	0.320	0.371	0.401	0.773	1.5
6	0.319	0.363	0.461	0.824	5.2
7	0.319	0.359	0.560	0.919	7.8
8	0.312	0.387	0.545	0.932	4.2
9	0.307	0.364	0.453	0.816	2.9
10	0.305	0.358	0.565	0.922	6.4
11	0.305	0.360	0.454	0.813	2.7
12	0.302	0.352	0.526	0.877	6.1
13	0.297	0.407	0.578	0.985	6.9
14	0.295	0.353	0.405	0.757	1.9
15	0.295	0.351	0.482	0.833	1.5
16	0.295	0.337	0.520	0.858	1.7

Remember that regression is just a fancy name for prediction. I will do the exact same steps that I did while doing regression in Chapter 17. I will graph each predictor separately. The first graph (on the next page) shows AVG as a predictor for WAR and the resulting S (RMSE, that is) = 1.63. S shows up on the worksheet AND the graph, so it is easy to find. But I chose Stat, Regression, Fitted line plot in Minitab. I put WAR as the Response, and AVG as the Predictor. And then OK.

I have to do this four times, because I have four predictors. I have to find S for each one. Each time I enter the information for the graph, I do not change WAR. WAR is my response for all four graphs. But I change my predictor each time.

Here is the graph for AVG and WAR:

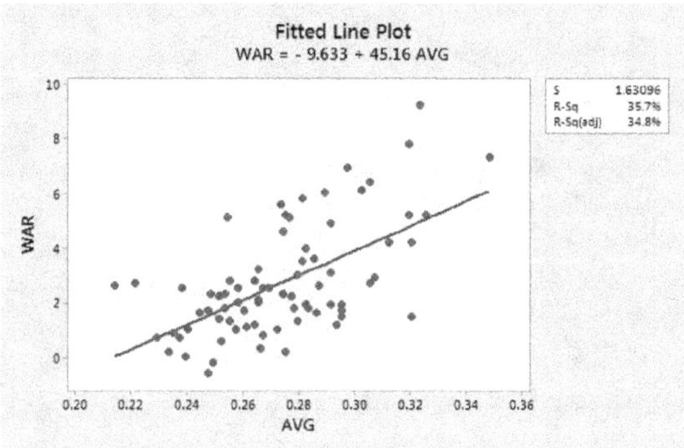

And the graph for OBP and WAR:

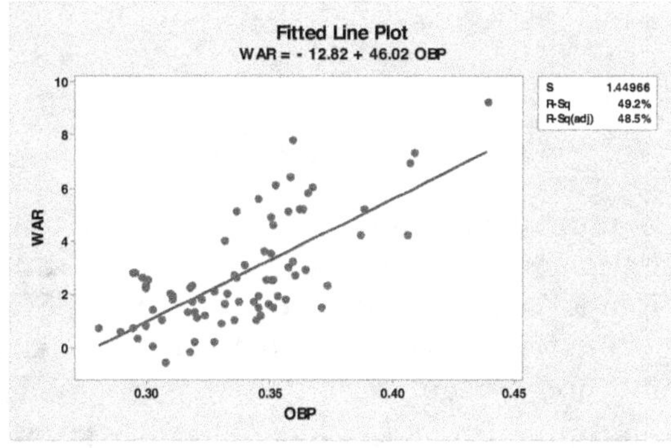

And SLG and WAR:

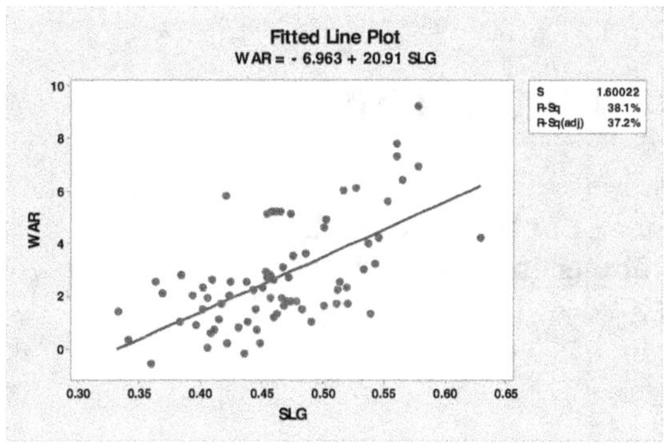

And OPS and WAR:

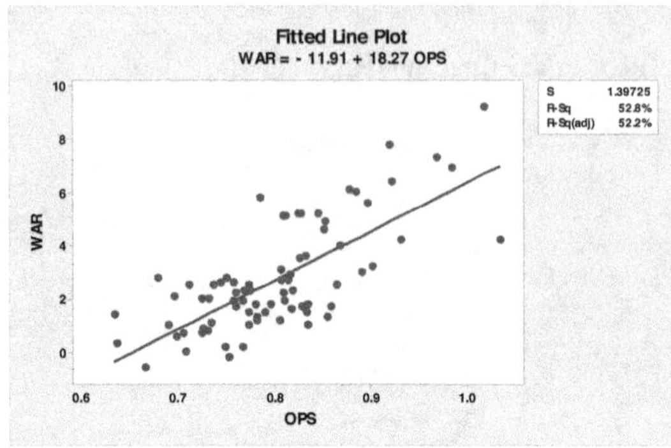

To summarize, here are the four values of RMSE (S) from the four graphs I just created:

Predictor	RMSE
AVG	1.631
OBP	1.445
SLG	1.600
OPS	1.397

The RMSE is much higher for AVG. This tells us that the residuals are much larger for AVG than any of the other three predictors. Remember the E in RMSE stands for error. A large RMSE means a large error. We are looking for the best predictor. We are looking for the least error. It is certainly not average.

The **best predictor** will always be the **lowest RMSE**. We want to minimize error, and the lower the number, the lower the error. OPS is the best predictor of WAR (at least in the AL in 2016) because it has the lowest RMSE.

PRACTICE PROBLEMS

Switching to volleyball, the men's team has been ranked number 1 in the country since the preseason. The team is led by another freshman phenom, Eli Gabriel Irizarry Pares, who has totaled an astounding 378 assists, and the season is not over yet. So far he has led the team to a 23-1 record, losing only 6 sets all year!

We will investigate men's volleyball to determine which is the best predictor of points (PTS): kills (K), total attacks (TA), assists (A), digs (DIGS), or total blocks (TOT).

Go to Springfield College Men's Volleyball, click on More, then Statistics, then Lineup. Copy the 6 columns (K, TA, A, DIGS, TOT, PTS) into Minitab. Use RMSE to decide which of the five variables (K, TA, A, DIGS, TOT) is the best predictor of points (PTS).

---

2 The Celtics are 9-2, the best record in the entire NBA. The number one scorer is Kyrie Irving, and he is only 18th in the league. We will use NBA scoring to answer this question.

Google "NBA stats", then click on the ESPN link. Under Offensive Leaders (the first box), click Complete leaders. Up top, change season to 2016-17, change League to East, change Position to Point guard. You should have 26 players. Copy five columns into Minitab for all 26 players: MPG (minutes per game), PTS (points), FG% (field goal %), 3P% (3 point %), and FT% (free throw %).

Then use RMSE (S on Minitab) to decide which of the four variables (MPG, FG%, 3P%, or FT%) is the better predictor of Points.

Borda count

Close don't count in baseball. Close only counts in horseshoes and grenades.

— *Frank Robinson, All-Star outfielder*

PREVIEW: ALL MVP AWARDS, HEISMAN TROPHIES, TOP 25 POLLS USE THIS METHOD

Borda count is an election method. Points are assigned based on how a voter ranks them. Candidates get one point for last place, two points for second-to-last, three points for third-to-last, and so on. The point values for all ballots are totaled. The candidate with the most points is the winner. Because it sometimes elects broadly acceptable candidates, rather than those preferred by the majority, Borda count is often described as a consensus-based electoral system, rather than a majoritarian one.

The Borda count method was developed independently several times but is named for the 18th-century French mathematician and political scientist Jean-Charles de Borda, who devised the system in 1770.

Borda count, and points-based systems similar to it, are often used to determine awards in competitions. Borda count is a popular method for determining awards for sports in the United States. It is used in determining the Most Valuable Player and Cy Young Award Winner in Major League Baseball; by the Associated Press and United Press International to rank teams in NCAA sports; to determine the winner of the Heisman Trophy; and so on.

Let's look at college football, for example. The Associated Press poll is a poll of football writers across the country. Each eligible voter — there are 60 of them — receives a ballot. Each Sunday, they fill out who they think are the top 25 teams in order of preference. They submit the ballot to the AP who awards the 25th ranked team on each ballot one point, the 24th ranked team two points, and so on up to the top-ranked team who gets 25 points. The AP then adds up all the points and publishes their rankings every Monday. Here's the poll from July 27, 2020:

The Associated Press Top 25 Poll
July 27, 2020

Rank	Team	Record	Points	Previous
1	LSU	15-0	1,550 (62)	1
2	Clemson	14-1	1,487	3
3	Ohio State	13-1	1,426	2
4	Georgia	12-2	1,336	5
5	Oregon	12-2	1,249	7
6	Florida	11-2	1,211	6
7	Oklahoma	12-2	1,179	4
8	Alabama	11-2	1,159	9
9	Penn State	11-2	1,038	13
10	Minnesota	11-2	952	16
11	Wisconsin	10-4	883	11
12	Notre Dame	11-2	879	14
13	Baylor	11-3	827	8
14	Auburn	9-4	726	9
15	Iowa	10-3	699	19
16	Utah	11-3	543	12
17	Memphis	12-2	528	15
18	Michigan	9-4	468	17
19	Appalachian State	13-1	466	20
20	Navy	11-2	415	21
21	Cincinnati	11-3	343	23
22	Air Force	11-2	209	24
23	Boise State	12-2	188	18
24	UCF	10-3	78	NR
25	Texas	8-5	69	NR

Borda count is a fun little diversion as we finish Part 2 of the course. Now, when you look at any top 25 polls, you know where those numbers come from.

CHAPTER E2

Practice exams for Part 2

SPRINGFIELD COLLEGE

Sports Statistics  Exam 2 Spring 2020

1 Lebron James was drafted in 2003 by the Cleveland Cavaliers. He immediately became one of the best players in the NBA. This is now his 17th season, and once again he is leading his Lakers to the NBA championship. They will beat Milwaukee in 6.

But as good a scorer as Lebron is, now at age 35 he has changed his game and he leads the NBA in .... assists! Let's graph his assists for his career to determine his trend throughout his career.

Google "Lebron James stats" and click on the ESPN link. Copy the column labeled AST (assists) into Minitab. It is the 6th column from the end. Copy his whole career (all 17 seasons), including 2019-20, but not the career total row at the end.

Last year UMass hockey reached their first-ever Frozen Four and a berth in the NCAA Division I National Championship game in which the Minutemen ultimately lost 3-0 to Minnesota-Duluth. The team posted a new program record for wins (31) while Cale Makar won the school's first Hobey Baker Award (college hockey's version of the Heisman).

UMass has never been this good in hockey. With Division I football as well, how has this affected how well UMass meets Prong 1 of the NCAA's Title IX requirements? The current athlete count at UMass is 266 for the women and 334 for the men. The undergraduate proportions are 49.8% women (enter .498 in Minitab) and 50.2% men (enter .502).

Do a hypothesis test (list H_o and H_A) to find your test statistic and p-value. Then type in your statistical conclusion and interpretation. Based on your hypothesis test, does UMass satisfy Prong 1 of Title IX?

Let's investigate Red Sox hitters in 2019. Google "Red Sox stats" and click on the MLB link. We will use the last 4 columns, AVG, OBP, SLG, and OPS. Scroll down and copy the first 22 Red Sox, down to Dustin Pedroia, into Minitab.

Then answer the question, Which of the first 3 columns (AVG, OBP, and SLG) is the better predictor of OPS? Use RMSE to decide and list the three S values in your answer.

Using the same data in question 3, we will investigate to determine if there is a relationship between slugging (SLG) and on-base percentage + slugging (OPS) in 2019 for the Boston Red Sox.

Graph

1. **Create a fitted line plot**.
2. **Describe the relationship.** (On the exam, put this in the footnote and print out the graph. (Don't forget your name!))

Correlation

3. **Calculate the correlation r.**
4. **Interpret r.**
5. **Calculate r².**
6. **Interpret r².**

Regression

7. **What is the regression equation?**

8. **Make a first prediction.** Mike Trout led the AL in slugging in 2019 at .645. If the Red Sox had signed him for 2019, what would you predict his OPS would have been for the Sox?

9. **Are you comfortable with this prediction and why?**

10. **Make a second prediction.** We traded Yoan Moncada to the White Sox in the Chris Sale trade. He has done very well and slugged .548 in 2019. If we had kept him, what would you predict his OPS would have been with the Sox?

11. **Are you comfortable with this prediction and why?**

12. **Make a third prediction.** Mookie Betts led the AL in 2018 in slugging but fell to 34th in 2019 with a .524 slugging percentage. What would you predict his OPS should have been?

13. **Are you comfortable with this prediction and why?**

14. **What is his residual?**

15. **Interpret his residual.**

SPRINGFIELD COLLEGE
Sports Statistics 🏈 Exam 2 Fall 2019

1 Alex Ovechkin, of the Washington Capitals, has been one of the best scorers over the last 15 years in the NHL. Let's graph his goals scored for his career to determine his trend.

Search Alex Ovechkin stats, and click on the NHL link. Scroll down to his career, and copy the fourth column for goals scored (G) into Minitab. Then graph his goals scored to determine his scoring trend throughout his career.

2 Louisiana State University will win the 2020 Football championship. But did they sacrifice Title IX to build their program? Let's investigate. LSU has 233 female athletes and 262 male athletes. Their undergrad proportion is 53% (.53) female and 47% (.47) male.

Do a hypothesis test (list H_o and H_A) to find your test statistic and *p*-value. Then type in your statistical conclusion and interpretation. Based on your test, does LSU satisfy Prong 1 of Title IX?

3 ESPN's version of the quarterback rating for basketball is the PER (Player Efficiency Rating). The stats we used to evaluate a player in the past were PTS (points), REB (rebounds), and AST (assists). Which of those three stats is the best predictor of PER?

Google "Celtics stats" and click on the ESPN link. Scroll down and copy four columns into Minitab: PTS, REB, AST, and PER. Copy for all players but do not include the total row.

Then we can answer the question, Which of the three columns (PTS, REB, or AST) is the best predictor of PER? Use RMSE to decide and tell me your three S values in your answer.

4 Using the same data in Question 3, we will determine if there is a relationship between rebounds (REB) and player efficiency rating (PER) for the 2019-20 Boston Celtics.

Graph

1. **Create a fitted line plot**.
2. **Describe the relationship.** (On the exam, put this in the footnote and print out the graph. (Don't forget your name!))

Correlation

3. **Calculate the correlation r.**
4. **Interpret r**.
5. **Calculate r^2.**
6. **Interpret r^2.**

Regression

7. **What is the regression equation**?
8. **Make a first prediction.** The Celtics were interested in Anthony Davis, but his 9.4 rebounds per game ended up with the Lakers. If he were playing for the Celtics, what would you predict his PER would be?
9. **Are you comfortable with this prediction and why**?
10. **Make a second prediction.** Al Horford was a good player with the Celtics, but is now in Philly and is averaging 6.8 rebounds per game. If he still played for the Celtics, what would you predict his PER would be?
11. **Are you comfortable with this prediction and why**?
12. **Make a third prediction**. The Celtics leading rebounder this year is Enes Kanter, with 7.7 rebounds per game. Based on the regression model, what would you predict his PER would be?
13. **Are you comfortable with this prediction and why**?
14. **What is his residual**?
15. **Interpret his residual**.

SPRINGFIELD COLLEGE
Sports Statistics ⚾ Exam 2 Spring 2019

1 Zdeno Chara remains one of the NHL's best defensemen, if not the most intimidating, even though he is now 41 years old. To celebrate that accomplishment, we will look up his average time on ice (ATOI) statistic to see if he has been able to maintain his ice time as he has gotten older.

Search "Zdeno Chara stats" click on the Hockey-reference link. Scroll down to ice time. Copy his career ATOI column beginning with 1998 into Minitab and then graph his ATOI and determine his trend throughout his career.

2 In January, Clemson won the college football national championship by destroying #1 Alabama, 44-16, and that was after they beat #3 Notre Dame in the semis, 30-3. They are now the first 15-0 team in 2 centuries! With the Tigers' focus on football, how do they treat their women's teams? Let's see if they satisfy Prong 1 of Title IX.

Clemson has 256 male athletes, and 254 female athletes (women's rowing is huge at Clemson, offsetting football). Their undergraduate proportion for men is 52% (.52), and for women, 48% (.48).

Do a hypothesis test (list H_o and H_A) to find your test statistic and *p*-value. Then type in your statistical conclusion and interpretation. Based on your test, does Clemson satisfy Prong 1 of Title IX?

3 You may have heard that the Red Sox won the 2018 World Series. That was after losing their first game of the season. Yesterday they lost the first game of the 2019 season to Seattle 12-4. Hopefully the Sox will follow that same blueprint and become the first team to repeat as World Champions since that team from NY did it.

We will look at pitching to see which is the best predictor of earned run average (ERA) for the 2018 Red Sox: strikeouts (SO), batting average against (AVG) or walks and hits per innings pitched (WHIP).

Search: Red Sox stats, click on the MLB link. There are two changes. In the second row, click on **PITCHING** and then below that click on **2018**. Copy ERA and the last 3 columns: SO, AVG, WHIP (include all 23 pitchers) into Minitab. Then use RMSE to decide which of the 3 variables

(SO, AVG and WHIP) is the best predictor of ERA. Be sure to list the three S values in your answer.

Using the same data in question 3, we will investigate to determine if there is a relationship between WHIP and ERA for the 2018 Boston Red Sox.

Graph

1. **Create a fitted line plot**.
2. **Describe the relationship.** (On the exam, put this in the footnote and print out the graph. (Don't forget your name!))

Correlation

3. **Calculate the correlation r**.
4. **Interpret r**.
5. **Calculate r^2**.
6. **Interpret r^2**.

Regression

7. **What is the regression equation**?
8. **Make a first prediction.** Oliver Perez led the Cleveland Indians in 2018 with a WHIP of .74. If the Red Sox had acquired Perez this year, what would you predict his ERA would be for the Red Sox?
9. **Are you comfortable with this prediction and why**?

10. **Make a second prediction**. Edwin Diaz was the best reliever in MLB in 2018. He led the majors in saves by a landslide. He also had an incredible .79 WHIP. What would you have predicted for his ERA if the Red Sox had acquired him from NY?

11. **Are you comfortable with this prediction and why**?

12. **Make a third prediction**. Craig Kimbrel was our closer in 2018, and had a WHIP of .99. What do you predict his ERA would have been using the linear model you constructed?

13. **Are you comfortable with this prediction and why**?

14. **What is his residual**?

15. **Interpret his residual**.

PART 3
Statistical Inference

We can't win at home. We can't win on the road. As general manager, I just can't figure out where else to play.

— *Pat Williams, Orlando Magic manager*

RYON HEALY POST-GAME INTERVIEW...

The last part of the book covers **statistical inference**. Here is where we stand so far:

In Part 1 we covered descriptive statistics which **describes** (summarizes) what things look like right now.

In Part 2 we covered predictive statistics which helps us guess — **predict** — what might happen next.

Now, in Part 3 we will use data to deduce — **infer** — what is going on. Inference uses reasoning to draw a conclusion based on evidence. Just the way a detective does.

We will use inference to make a statistical estimate based on a sample. For example, we might want to know how many home runs JD Martinez might hit in 2020. To make that prediction, we would use the home runs he hit each season of his career. Here is our sample of JD's home runs hit per season:

Year	JD Martinez home runs
2011	6
2012	11
2013	7
2014	23
2015	38
2016	22
2017	45
2018	43
2019	36

We use JD's home runs from the past to predict how many home runs JD will hit in 2020. We will *infer* from our sample. That is **statistical inference**. At the end of Chapter 22, I will use JD's home run sample to estimate how many home runs he will hit in 2020.

Each chapter in Part 3 has an optional section at the end that delves deeper into the statistics of that particular chapter.

Tools for statistical inference

I don't know whether you know it, but baseball's appeal is decimal points. No other sport relies as totally on continuity, statistics, orderliness of these. Baseball fans pay more attention to numbers than CPAs.

— *Jim Murray, sportswriter*

PREVIEW: BUILDING OUR TOOLBOX

There are two methods of statistical inference. In statistics we call the methods "tests". Statistical tests. The two tests are **hypothesis testing** and **confidence intervals**. Hypothesis tests are the same hypothesis tests we have done in Part 1 and Part 2 of the book. Confidence intervals are new. Before we dive into the methods of inference, we will talk about confidence intervals. Then you will have both tools in your toolbox as we learn inference.

Confidence intervals

I decided to do a bike ride down to East Hartland from my home in Westfield. Based on my previous rides (my sample), I tell my wife I estimate I will be gone 3 hours. In statistics, this is my **point estimate**. A point estimate is a "best guess" or "best estimate" based on the average of my previous rides.

I feel comfortable telling my wife it will take 3 hours, because I have done that ride many times. I have a large sample size. But there are a lot of variables in a ride, like wind, heat, traffic, fueling and caffeine, previous day's workout or lack thereof, and mechanical issues with the bike. The probability of me taking exactly 3 hours is very low. As a matter of fact, it would be surprising to roll back home in EXACTLY 3 hours!

To be safe, a better answer is to tell her: 3 hours, give or take 1 hour. Or 3 hours, +/- 1 hour. The +/- in statistics is known as the margin of error. Utilizing the margin of error means I will be back in 2 to 4 hours. I did 3 − 1 = 2 (my low end of the interval) and 3 + 1 = 4 (my high end of the interval). I feel much better that I could come home anytime in that range of 2 to 4 hours, rather than exactly 3 hours.

That range in statistics we call an interval. And since I made the interval pretty wide (2 to 4 hours on a 3 hour ride) I am pretty comfortable I will make that interval. That comfort is reflected in my confidence. I am pretty confident. 100% confident? No, I may break a chain, or get lost, and saunter back home in 4½ hours. It hasn't happened, but it could. My confidence level is closer to 95%. In stats, we call that a 95% confidence interval.

Here is how we write my 95% confidence interval: (2, 4). My statistical conclusion is: I am 95% confident that I will be back home somewhere in between 2 to 4 hours. Anytime between 2 and 4 hours is therefore expected, it is just random variation. If I come in outside of the 95% confidence interval, like 1:45 (or any time below 2 hours), or 4:30 (or any time above 4 hours), it is not random variation. Something is going on. I got lost. My chain broke. Or with the 1:45 I was over-caffeinated and went wicked fast!

In the real world, everyone has a different confidence level. Mine was 95% here, but maybe you are 90%, or 99%, or any value. 90% confidence is used for fantasy football owners, not life or death, right? And 99% confidence levels are for MLB owners signing players for $30 million guaranteed for 4 years. More important to get that right, you want to be very confident.

In sports statistics, we will use the middle ground for every problem. Every confidence interval we do, we will do it at the 95% confidence level, to keep it consistent for you guys. It matches the .05 we have been using in hypothesis testing. .05 is 5%. If we are lower than 5%, we reject the null hypothesis.

If we are in the 95% confidence interval, it is just random variation. Being outside the 95% confidence interval means it is not random variation. What is outside a 95% confidence interval? 100% - 95% = 5% = .05! This way our two tests test the sample at the same 5% level.

Proportions vs. means

A proportion is a fraction, a decimal, or a percent. In the problems coming up we will recognize it with a P. P for proportion problems. A mean is an average. In the problems coming up, we will use M for mean. P for proportion, M for mean. They are both descriptive statistics. We can summarize ANY data in sports with either a proportion or a mean. Then after we summarize the data, we will test the data to determine if something is going on with the data.

If something is going on, we will find the sample is statistically different. In statistical terms, we reject the H_0. If nothing is going on, it is random variation, and we fail to reject H_0.

> **Proportion Fun Fact**
>
> As of the summer of 2020, the NBA's all-time leader in free-throw percentage is Stephen Curry. Steph has made 2560 free-throws out of 2827 attempts, for a 90.6% success rate.
>
> Compared to Shaquille O'Neil, who made 5935 free throws out of 11252 attempts, or an astonishingly low 52.7%. If Shaq had hit the same percentage as Steph, Shaq would have had 10189 free throws made, instead of the 5935. That would be an additional 4254 points in his career!

Here is a proportion example with Patrice Bergeron. Every year he leads the Bruins in face-off wins. He did it again in 2019-20. In the regular season, he won 758 of the 1311 face-offs he took. Why is this a proportion example? Because we can write the data as a fraction, a decimal, or a percent. Bergeron won 758/1311 (fraction), .57818 (decimal gotten by dividing), and 57.82% (percent gotten by moving the decimal two places to the right) of his face-offs.

Now I can analyze the data and note that 57.82% in the NHL is very high! It is very hard to win that many face-offs. Of all the Bruins in 2019-20, only one other Bruin broke 50%. And that was David Krejci at 50.34%. Everyone else was less than half. But for Bergeron, his proportion, his P, is equal to .5782. Or P = .5782.

How about means? Well, on our second day of class, we played mini-putt and recorded our scores for the front nine and the back nine. Then to compare our performance, we did an average on Minitab. Average we now call the mean. Now, here we are at the end of the semester, and when we have numbers, the mean is still the best way to go!

Here are some numbers for Patrice Bergeron:

18:31   16:47   18:33   17:07   16:10   18:08   21:25   21:38   21:20

Notice, no fractions, no decimals, no percents. Therefore, no proportion. We use the mean (M). What are these numbers? They are Bergeron's time on ice for the first 9 games of the 2019-20 season. Based on our mini-putt exercise, we want to find the mean for Bergeron's ice time. In Minitab I put the ice time for all 61 games Bergeron played (not just the first nine games that you see above) into C1, and calculated his mean. His mean ice time for 2019-20 was 18:44 minutes per game.

To summarize, a proportion (P) is a fraction, a decimal, or a percent. A mean (M) is an average of numbers. And those are the only descriptive statistics you need to know for the rest of the course! Every problem from today until the final exam is either a proportion (P) problem or a mean (M) problem.

Mean Fun Fact

Wilt Chamberlain had a career that will never be duplicated. He is the only NBA player to score 100 points in a game. But he was not a one game wonder. He played 1045 games over 14 seasons, and he scored an average (mean) of 30.1 points per game! Not just one game, he averaged over 30 points per game for his CAREER.

While visiting with Wilt, we notice he also rebounded very well. For his CAREER again, he averaged (mean) 22.9 rebounds per game. 22.9 rebounds night after night, for 1045 games. He is the only NBA player to average 20+ rebounds and 30+ points for EVERY game in that season. And Wilt did that for seven different seasons!

Finally, in his third season, he averaged (mean) 48.5 minutes per game. I had to check to make sure in the 1960's that NBA games were only 48 minutes long. They were! Wilt basically played every minute of every game, including OT. And last fact, during that season, his mean scoring per game was 50.4 points. Yes, 50.4 points EVERY game!

Summary of statistical inference tests

In Part 3 of this book, we are learning how to do statistical inference. There are five methods of inference we will be doing. As we get to these methods in the book, you will learn about them in detail. But here is a quick summary for you of what lies ahead:

1-Proportion (P): This is for a problem with one fraction, decimal or percent. Tom Brady completed 373 out of 613 passes in 2019. That can be written as 373/613, or .60848, or 60.85% (a fraction, decimal or percent). (We'll look at Tom's performance at the beginning of **Chapter 21**.) It is a 1-proportion problem as there is only one fraction, or decimal, or percent. Compare this to the 2-proportion below.

1-Sample Mean (M): Above, we see 1-proportion. What if I had JD Martinez's home runs every season of his career? I looked them up and here they are:

6   11   7   23   38   22   45   43   36

This is different from Tom Brady. With TB12 we had a proportion. Here with JD we have numbers. His summary is an average, or a mean. A quick summary of JD's career is a mean (it was 25.67 home runs per season). And we have just one player (as opposed to the 2-sample mean below), so JD's home runs is an example of a 1-sample mean question. We will do JD at the end of **Chapter 22**.

2-Sample Mean (M): In 2020, the SC women's lax team scored the following number of goals: 18, 11, 12, 1, 4. The SC men's lax team scored the following number of goals: 10, 8, 6, 9.

We have numbers, like JD above. That makes it a sample-mean problem. The reason it is a 2-sample mean problem specifically is we have two samples! The women's lax team and the men's lax team. We'll look at them as a guided example in **Chapter 23**.

2-Proportions (P): In the summer of 2015 the NFL moved the extra point back 13 yards. This was because year in and year out, kickers were almost automatic from the 2 yard line. But how would they do from the 15 yard line?

I looked up the last year of the old distance (2014) and I looked up 2019 to see if anything had changed. In 2014, kickers made 1222 kicks out of 1230 attempts. Last year they made 1136 kicks out of 1210 attempts.

I found 1222/1230, = .99349 = 99.35%. AND, I found for last year, 1136/1210 = .93884 = 93.88%. That is two separate proportions, therefore we do a 2-proportion test. We'll investigate the NFL kickers in a guided example in **Chapter 24**.

ANOVA (M): With ANOVA we can do 3 or more samples of means, for instance we can compare several different seasons for LSU.

Was the 2020 National Champion LSU Tigers the best team ever? It is arguable. But, I will settle for the best LSU team over the last three years. I looked up the Tigers total yards of offense for the last three years, and listed them below. After we enter the data in Minitab, we will use ANOVA to answer the question, Is there a statistical difference in yards gained for LSU in 2017, 2018 and 2019? We'll see **Chapter 25**.

2019 LSU Tigers offense	2018 LSU Tigers offense	2017 LSU Tigers offense
472	296	479
573	335	454
611	370	270
599	409	414
601	573	428
511	372	341
413	475	363
508	239	593
559	196	306
716	359	415
612	552	281
553	496	601
481	555	399
693		
631		

Test statistics

The **test statistic** is a value we calculate when performing a hypothesis test. We did hypothesis testing in Part 1 of this book when we did the distribution of athletes on our Springfield College varsity rosters to see how well they fit the numbers we expected. We used the **chi-square test statistic** to answer that question.

In Part 2, we did Title IX and we investigated whether a college violated Prong 1 of Title IX. To answer that question, we did a hypothesis test, again utilizing the chi-square value as our test statistic.

In this last section of the book (Part 3), we will continue to do the hypothesis tests, and add a second test called a 95% confidence interval. For the hypothesis testing, we will perform the same six steps we learned back in Part 1, but we will finally change the test statistic. No more chi-square.

The first test statistic we do is called z. Yes, the same z we learned with the normal approximation. The same z of z-score fame. We know z, we may even like z. We are at least familiar with z. And it is easy to find. In stats, we have a table of z values. But for us, Minitab will calculate it for us, no need to look it up. We will use z for our test statistic when we do proportions. Proportions use z. Proportions use z.

If we like z that much, why can't we use it with means? Well, statistically, it is because we do not know the population standard deviation. We need that for z. What do we do, punt? No, we have a real good substitute test statistic called t. t looks like z, same normal looking distribution, it is only off by one degree of freedom. In other words, it is very close to z, and it works! We will use t when we do 1-sample mean and 2-sample means. Means use t. Means use t.

What if we have 3 or more sample means? Oh, that is our last chapter! We will use a statistical test called analysis of variance (ANOVA). The test statistic for ANOVA is F. ANOVA is F. ANOVA is F.

Test	Test Statistic
1 Proportion	z
2 Proportions	z
1-Sample Mean	t
2-Sample Means	t
ANOVA	F

Here are all the menu choices on Minitab.

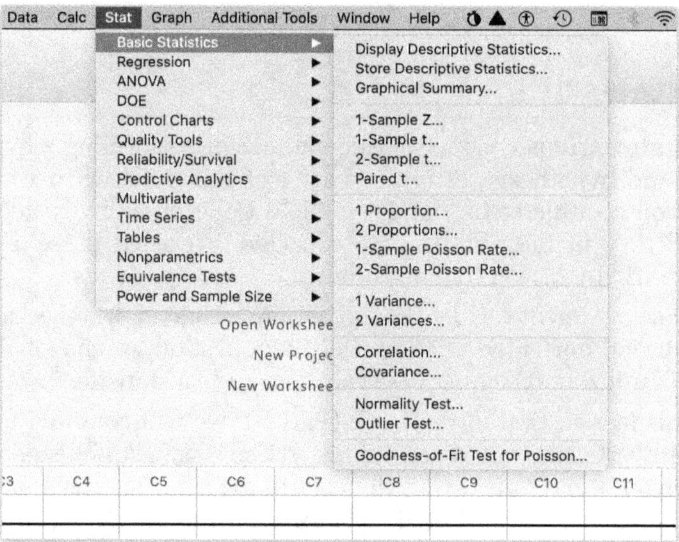

In summary:

To get the ...	On Minitab choose ...		
t-test statistic for the 1-Sample Mean test	Stat then	Basic Statistics then	1-Sample t...
t-test statistics for the 2-Sample Means	Stat then	Basic Statistics then	2-Sample t...
z-test statistic for the 1-Proportion test	Stat then	Basic Statistics then	1 Proportion...
z-test statistic for the 2-Proportion test	Stat then	Basic Statistics then	2 Proportions...
F-test statistic for the ANOVA test	Stat then	ANOVA	One-way...

10-step answer format

All questions dealing with statistical inference will use the same **10-step answer format**. The first six steps are familiar to us, they are what we called the hypothesis test. In statistical inference, the hypothesis test will test the SAMPLE so we can make an inference about the POPULATION.

In statistical inference, we want to be absolutely sure of our conclusions. Therefore we add a second test called a 95% confidence interval. The 95% confidence interval will test the same SAMPLE as in the hypothesis test, and allow us to make a second inference about the POPULATION.

Finally, in Step 10 of the 10-step process, we look to see if the 95% confidence interval we just did CONFIRMS the hypothesis test. Do the two tests give us the same result? Do the two tests confirm each other? The same result is a confirmation of the other test. Same result is confirm.

Here are the ten steps of the **10-step answer format** with my notes:

Test 1: Hypothesis Test

1. **H_0:**
   This is the null hypothesis, where the = sign goes. Because null means nothing is going on here, we are equal to the old number. P = the old number. M = the old number. In the 2-proportion or 2-sample means, the two proportions are equal ($P_1 = P_2$) or the two means are equal ($M_1 = M_2$).

2. **H_A:**
   This is the alternative hypothesis, where the inequality goes. We will use less than (<) or greater than (>) depending on the wording in the problem. Are they worse? Use <. Have they improved? Use >. Only ANOVA will be different, using At least one mean is different.

3. **Test statistic and its value:**
   No more chi-square. Minitab will use either z, t, or F depending on the problem.

4. ***p*-value:**
   Same probability from before that we will compare to .05

5. **Statistical conclusion**
   Same as before. Either: Since the *p*-value < .05, I reject H_0 or Since the *p*-value > .05, I fail to reject H_0.

6. **Interpretation**
   If we reject, there is something going on. If we fail to reject, nothing is going on, it is just random variation.

Test 2: Confidence Interval

7. **95% Confidence interval**
   Just record the low and high ends of the interval here.

8. **Statistical conclusion**
   This one always starts: I am 95% confident.... What follows depends on the problem.

9. **Interpretation**
   When we have one sample or one proportion, look to see if our old number is in the confidence interval. If it is, nothing is going on. If it is outside the confidence interval, something is going on.

   If we have 2-proportions or 2-sample means, we will learn the sign rules. The sign rules are negative to negative, or positive to positive , means there is a difference. Negative to positive means there is NO difference. And with ANOVA we will look to see if the confidence intervals do not overlap (there is a statistical difference) or if they overlap (no statistical difference).

Confirmation

10. **Do the results of the two tests confirm each other?**
    Con means together, *firm* means strengthen. We are now doing two tests on the same data. *Together* do the two tests *strengthen* our conclusion? Together, we make our conclusion stronger.

    How do we do that? Well, 1-6 above is the old test, the Hypothesis test. 7-9 is the new test, the Confidence interval. Both test the same data. Do they have the same result? Look at 6 and 9 above. If 6 and 9 have the same result, then they confirm each other and you answer yes. If 6 and 9 above give you different answers, then they do not confirm each other, and you answer no.

10-step answer format without teaching notes

Taking away my notes, all statistical inference testing will follow this format:

Test 1: Hypothesis Test

1. **H_0:**
2. **H_A:**
3. **Test statistic and its value:**
4. ***p*-value:**
5. **Statistical conclusion:**
6. **Interpretation:**

Test 2: Confidence Interval

7. **95% Confidence interval:**
8. **Statistical conclusion:**
9. **Interpretation:**

Confirmation

10. **Do the results of the two tests confirm each other?**

1-proportion statistical test

Baseball is the only field of endeavor where a man can succeed three times out of ten and be considered a good performer.

— *Ted Williams, Red Sox legend, last player to hit over .400*

PREVIEW: ANSWER "BUT WAS IT STATISTICALLY WORSE?" WITH JUST ONE FRACTION

I will guide you through a Tom Brady example to learn Minitab for a 1-proportion test. Then I will guide you through the same Tom Brady question to learn how to do the **10-step answer format.** Then I will give you three questions to practice on your own. After the three practice questions, I will give you the answers so you can check yourself and make sure you are comfortable with 1 proportion. If you are not 100% comfortable, there is a bonus question, with the answers.

GUIDED EXAMPLE FOR 1-PROPORTION WITH TOM BRADY

Tom Brady had one of the worst seasons of his career in 2019, completing only 373 out of 613 passes. His career completion percentage is 63.8%. Perform a hypothesis test and a 95% confidence interval to answer the question, Was Tom Brady statistically **worse** in 2019?

We will need the test statistic, p-value, and the 95% confidence interval to answer that. Minitab to the rescue! Here are the step-by-step Minitab instructions for finding the test statistic, p-value, and the 95% confidence interval.

Choose Minitab options

Choose Stat from the menu bar then Basic Statistics. This opens all the basic statistics options. There are many, as you can see:

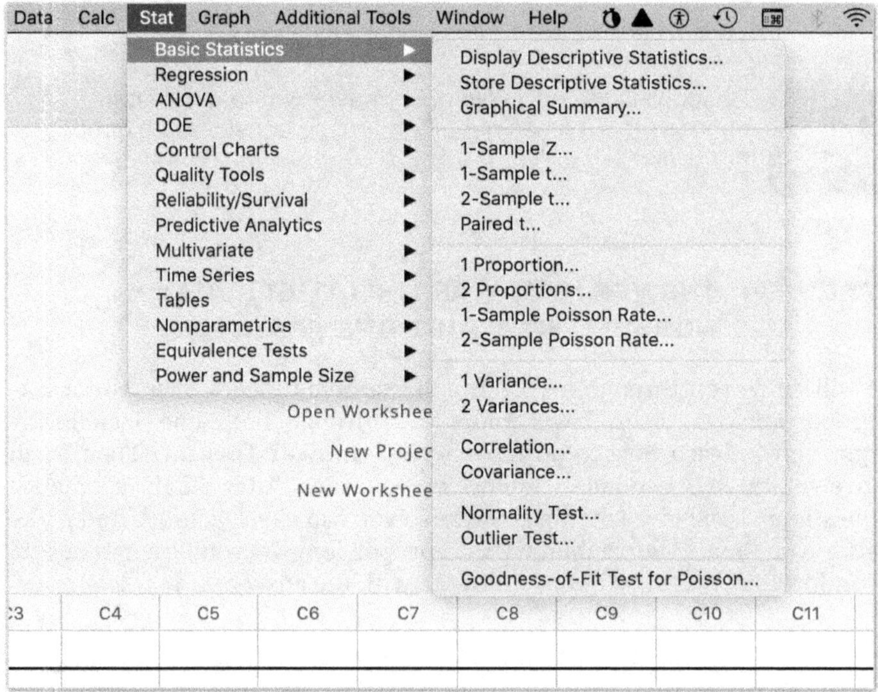

Which option do we pick? Well, the problem gives us 373 out of 613, which is a proportion. Why? Because we can write it as a fraction

(373/613), a decimal (.60848), or a percent (60.85%). Fraction, decimal, percent. Fraction, decimal, percent. Those are always proportion problems. In the context of this problem, Brady completed 60.85% of his passes in 2019.

If Brady's career numbers are 63.8%, and the 2019 season was 60.85%, wasn't he worse? We will find out there is a difference between worse and statistically worse. Yes, his completion percentage was down, but maybe it was random variation (injuries, no Gronk, etc.). If that was the case then Tom would not be statistically worse. We will learn how to test his numbers to determine if his decline was random or not.

There are two options that are labeled proportions: 1 Proportion and 2 Proportions. We have just TB12. One player, for one year. That is a 1-proportion problem. So choose 1 Proportion.

The screen defaults to One or more samples, each in a column. We have the **summary** of 2019 (373 out of 613). So choose **Summarized** data.

Now two boxes show up. Number of events and Number of trials. Trials is always the total (in this case, 613 total pass attempts). Events is the specific statistic we are testing, TB12's completion percentage. Enter 373 in the top box and 613 in the bottom box.

The next prompt is for doing a Hypothesis test. My question above read "Perform a hypothesis test", so yes we want to. We will always want to.

Check Perform hypothesis test.

That opens up the box to the right, the Hypothesized proportion. The hypothesized proportion is the number in the H_0. For us, the H_0 is no change, nothing going on. Tom Brady is still a 63.8% passer. TB12's proportion is still 63.8%. P = 63.8%. Unfortunately, Minitab does not accept percents, so move the decimal over two places and enter .638 in the box for Hypothesized proportion.

Then click Options. There are three. The top option, Confidence level defaults to 95. Good! Do not change it. My question above asked you to perform a 95% confidence interval.

Option two is Alternative hypothesis. The H_A defaults to ≠ (not equal to). But look at my question. I asked was TB12 statistically **worse**? Worse means <. Less than. The top choice. When we get there, H_A < .638!

The last option defaults to the method. Change Exact to Normal approximation. The same normal approximation we did with z-scores earlier.

Click OK, OK.

Done!

Here is a composite of the screens in Minitab for how to load the data and the options we used. The results of our test are in the worksheet on the left:

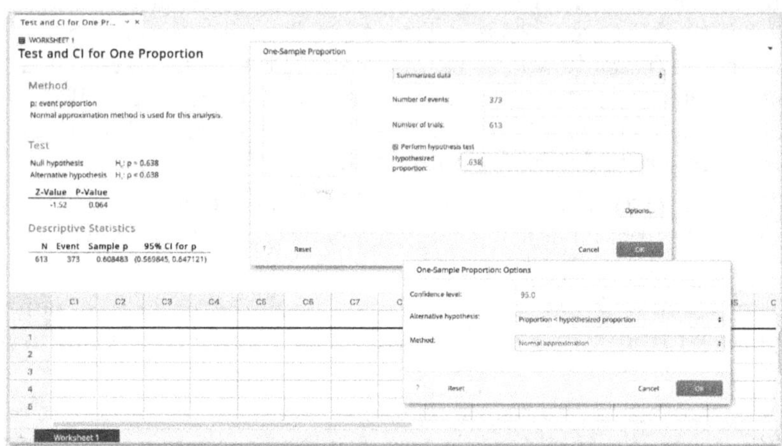

Notice the output above has your test statistic, *p*-value and 95% confidence interval. It did NOT originally have the 95% confidence interval. What I did to get the interval is go back to Minitab options and I changed the H_A to ≠. That is it! One change. In Minitab I chose Stat, Basic Statistics, 1 Proportion, Options, and here I changed the H_A only. Everything else stays the same. You will do that every test we do to get the confidence interval to show up.

Now we can do our guided example using the **10-step answer format** we learned in Chapter 20. We will use the test statistic (z = -1.52), *p*-value (.064) and 95% confidence interval (.569845, .647121) that Minitab just calculated for us as we answer the Tom Brady question.

Answer to guided example

Here is the guided example question again:

Tom Brady had one of the worst seasons of his career in 2019, completing only 373 out of 613 passes. His career completion percentage is 63.8%. Perform a hypothesis test and a 95% confidence interval to answer the question, Was Tom Brady statistically **worse** in 2019?

Test 1: Hypothesis Test

1. **H_0:** P = .638

   P stands for proportion, which is a fraction, decimal or percent. TB12 completed 373/613 passes, a fraction, so when you see fractions, or decimals, or percents in a problem, use P here. The null hypothesis is no change, no change for TB12 is that he still completes 63.8% of his passes, or P = .638

2. **H_A: P < .638**
   Keep the P, keep the .638, and we use .638 because Minitab does not accept the 63.8%. I choose < because the question asks is TB12 worse? Worse means his completion % is going down. The alternative is he is completing less than 63.8% of his passes.

3. **Test statistic and its value:** z = -1.52
   A negative test statistic is OK, it means he is lower. z is our new test statistic though. It comes from the normal distribution. Remember in the second exam we compared z-scores? It is that same z!

4. ***p*-value:** .064
   There is a 6.4% chance that Tom Brady is still a 63.8% QB, given his 2019 season. That is a low probability, but not low enough statistically.

5. **Statistical conclusion:** Since the *p*-value > .05, I fail to reject H_o. Same two choices in inference, > .05 is fail to reject, and < .05 is reject.

6. **Interpretation:** Tom Brady was NOT statistically worse in 2019, the difference we see is just random variation. He could still be a 63.8% QB for Tampa Bay.

Test 2: Confidence Interval

7. **95% Confidence interval:** (.569845, .647121)
   Tom Brady in 2019 was 373/613 = .60848. The .60848 is his point estimate, or the center of the confidence interval. Minitab took the .60848 and added and subtracted the margin of error to calculate Tom's 95% confidence interval.

8. **Statistical conclusion:** I am 95% confident that Tom Brady will complete between 57% and 65% of his passes.

   Notice I changed the decimals to percentages by moving the decimal two places to the right.

9. **Interpretation:** Since .638 is in the confidence interval, TB12 could still be a 63.8% quarterback. He is NOT statistically worse in 2019.

Look to see if the null hypothesis (the old number of .638) is in the interval. If it is in the confidence interval, he could still be that. You are 95% confident that he could still be that. But, if it is outside the interval, then he is different.

Confirmation

10. **Do the results of the two tests confirm each other?** Yes, the confidence interval and the hypothesis test both show that Tom was NOT statistically worse in 2019.

Look at parts 6 and 9. If they tell you the same thing (good or bad), they confirm. Same result = confirmation, yes. Different results mean the two tests do NOT confirm each other.

Answer without teaching notes

We are done! My guided example includes my notes. All you need to print out for the next three practice problems are the 10 steps of the **10-step answer format**, like this:

Test 1: Hypothesis Test

1. **H_0:** $P = .638$
2. **H_A:** $P < .638$
3. **Test statistic and its value:** $z = -1.52$
4. **p-value:** .064
5. **Statistical conclusion:** Since the p-value > .05, I fail to reject H_0
6. **Interpretation:** Tom Brady was NOT statistically worse in 2019, the difference we see is just random variation. He could still be a 63.8% QB for Tampa Bay.

Test 2: Confidence Interval

7. **95% Confidence interval:** (.569845, .647121)
8. **Statistical conclusion:** I am 95% confident that Tom Brady will complete between 57% and 65% of his passes.
9. **Interpretation:** Since .638 is in the confidence interval, TB12 could still be a 63.8% quarterback. He is NOT statistically worse in 2019.

Confirmation

10. **Do the results of the two tests confirm each other?** Yes, the confidence interval and the hypothesis test confirm each other. They both show that Tom was NOT statistically worse in 2019.

PRACTICE PROBLEMS

Please answer the three questions below using the **10-step answer format**. Try these three problems on your own, THEN check the answers to make sure we get the same values from Minitab. It is hard to not look at the answers first. For all of us, human nature is to look at the answers first and say, "Oh, that is what I would have said!" Practice that ingrains the methods and answers for us. The more we practice, the less test anxiety we have!

In the 2018-19 season, the Boston Celtics won .598 of their games under Kyrie Irving. Out went Kyrie, in came Kemba Walker. In 2019-20 season with Kemba, the Celtics won 43 games out of 64. Were the Celtics statistically **better** in 2019?

Jake Ross had a phenomenal basketball season his senior year at Springfield College. He is in the running for Division III basketball player of the year. This year (2019-20) Jake made 253 out of 554 shots he took. In 2018-19, his shooting percentage was 43.8%. Did Jake's shooting statistically **improve** in 2019-20?

Mookie Betts was one of the best all-around players the Red Sox ever had. He was MVP in 2018, while also winning the World Series. In 2018 he hit .346. In 2019 he "only" had 176 hits in 597 at bats. Was he statistically **worse** in 2019?

Mookie Betts Fun Fact

On August 13, 2020 in a game against the San Diego Padres, Mookie went 4-for-4. Three of those hits were home runs. The three homer game is special, and Mookie has now done it six times in his career. Obviously the first five games were with the Red Sox.

Mookie tied the major league record for three-homer games. The other two players are Sammy Sosa and Johnny Mize, both retired. Sosa needed 2364 games, Mize needed 1884 games, and Mookie? Just 813 games! And at 27, Mookie was the youngest of the trio when he hit his three homers against San Diego.

Answers to practice problems

In the 2018-19 season, the Boston Celtics won .598 of their games under Kyrie Irving. Out went Kyrie, in came Kemba Walker. In 2019-20 season with Kemba, the Celtics won 43 games out of 64. Were the Celtics statistically **better** in 2019?

Test 1: Hypothesis Test

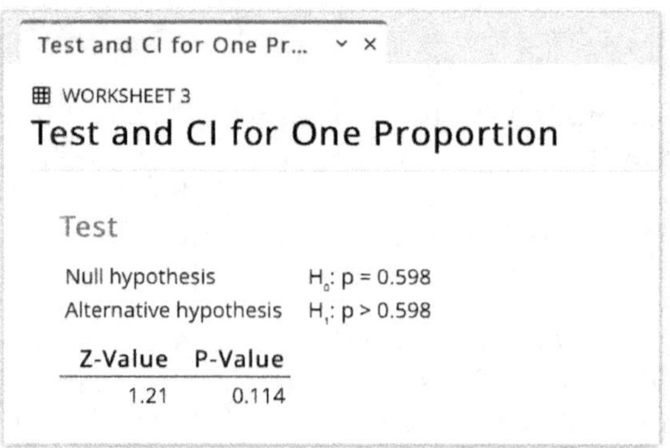

1. **H_0: P = .598**
2. **H_A: P > .598**
3. **Test statistic and its value:** z = 1.21
4. **p-value:** .114
5. **Statistical conclusion:** Since the p-value > .05, I fail to reject H_0
6. **Interpretation:** The Celtics are NOT statistically better in the 2019-20 season, the differences we see are just random variation.

Test 2: Confidence Interval

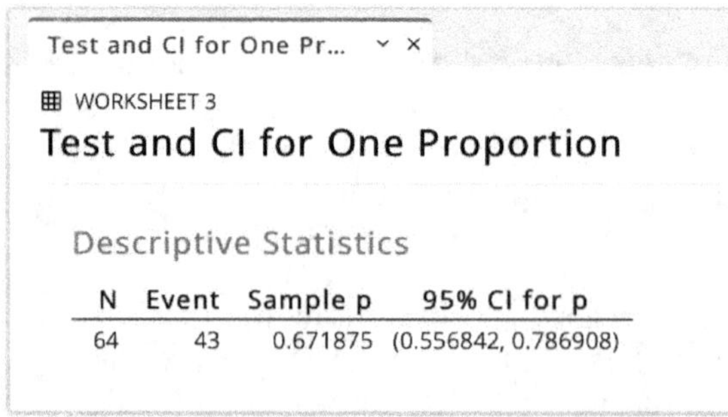

7. **95% Confidence interval:** (.556842, .786908)

8. **Statistical conclusion:** I am 95% confident the Celtics will win between 56% and 79% of their games with Kemba.

9. **Interpretation:** Since .598 is in the confidence interval, the Celtics could still be a .598 winning team. The Celtics are NOT statistically better.

Confirmation

10. **Do the results of the two tests confirm each other?** Yes, the confidence interval and the hypothesis test both show that the Celtics are NOT statistically better.

Jake Ross had a phenomenal basketball season his senior year at Springfield College. He is in the running for Division III basketball player of the year. This year (2019-20) Jake made 253 out of 554 shots he took. In 2018-19, his shooting percentage was 43.8%. Did Jake's shooting statistically **improve** in 2019-20?

Test 1: Hypothesis Test

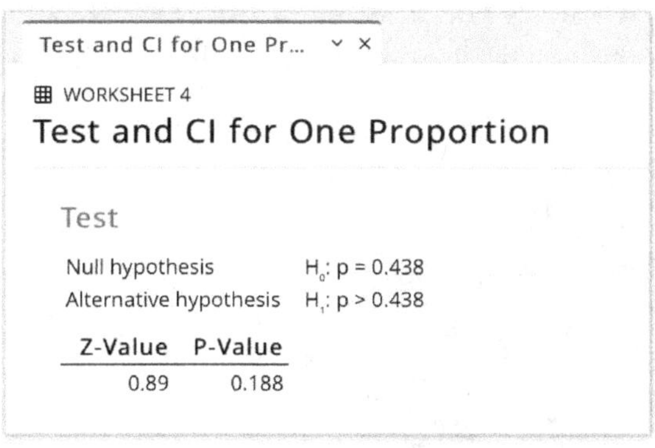

1. **H_0:** P = .438
2. **H_A:** P > .438
3. **Test statistic and its value:** z = .89
4. ***p*-value:** .188
5. **Statistical conclusion:** Since the *p*-value > .05, I fail to reject H_0
6. **Interpretation:** Jake did NOT statistically improve in 2019-20. The differences we see are just random variation.

Test 2: Confidence Interval

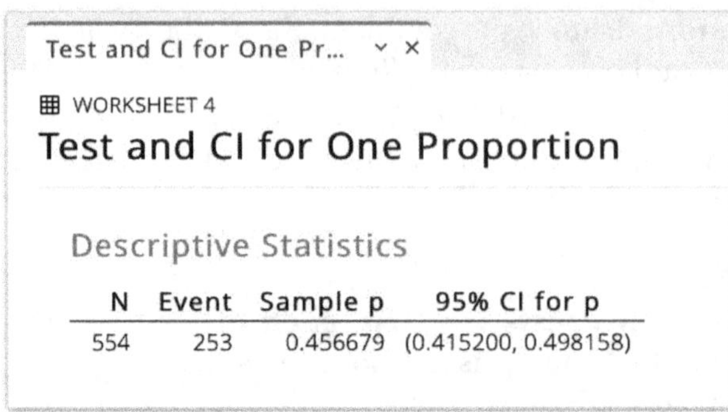

7. **95% Confidence interval:** (.415200, .498158)

8. **Statistical conclusion:** I am 95% confident that Jake will make between 42% and 50% of his shots in 2019-20.

9. **Interpretation:** Since .438 is in the confidence interval, Jake could still make 43.8% of his shots. Jake has NOT statistically improved in 2019-20.

Confirmation

10. **Do the results of the two tests confirm each other?** Yes, both the confidence interval and the hypothesis test show that Jake is NOT statistically better this year.

Mookie Betts was one of the best all-around players the Boston Red Sox ever had. He was MVP in 2018, while also winning the World Series. In 2018 he hit .346. In 2019 he "only" had 176 hits in 597 at bats. Was he statistically **worse** in 2019?

Test 1: Hypothesis Test

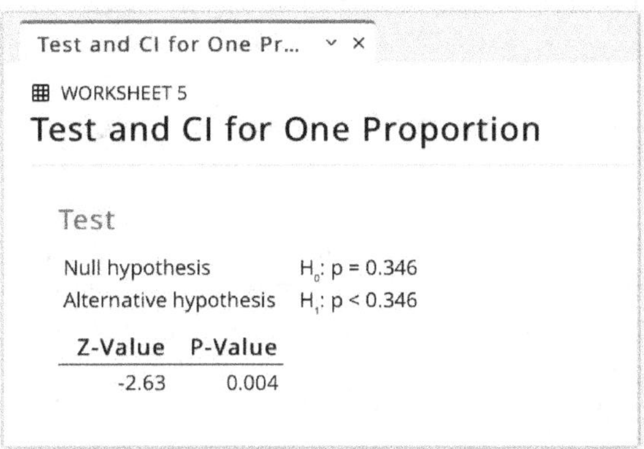

1. **H₀:** P = .346
2. **Hₐ:** P < .346
3. **Test statistic and its value:** z = -2.63
4. *p*-**value:** .004
5. **Statistical conclusion:** Since the *p*-value < .05, I reject H₀
6. **Interpretation:** Mookie was statistically worse in 2019.

Test 2: Confidence Interval

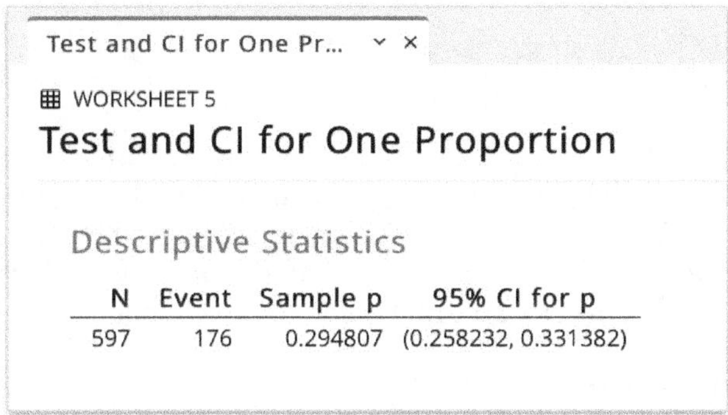

7. **95% Confidence interval:** (.258232, .331382)

8. **Statistical conclusion:** I am 95% confident that Mookie would hit between .258 and .331.

9. **Interpretation:** Since .346 is NOT in the confidence interval, and the confidence interval is below .346, Mookie is statistically worse in 2019

Confirmation

10. **Do the results of the two tests confirm each other?** Yes, the confidence interval and the hypothesis test both show that Mookie is statistically worse in 2019.

BONUS PROBLEM WITH SC'S JAKE ROSS

Jake Ross again (I know, it is Jake Ross all the time, but this time we will investigate his 3-point shooting) had a phenomenal senior season. He is in the running for the 2020 Division III basketball player of the year. This year Jake made 70 out of 180 3-point shots he took. In 2018-19, his shooting percentage was .308 (30.8%). Did Jake's shooting statistically **improve** in 2019-20?

Teaching note: 70 out of 180 = 70/180 = .38889 = 38.89% = 39%. If I had asked: Did Jake's 3 pt. shooting improve in 2019? You would have said, YES! Last year, about he shot about 31%. This year, about 39%! Improvement, right? And you are right. But, in stats we look at statistical change, not just change. We will test twice to see if Jake's numbers are statistically **better**, or if the differences we see are just random variation from one year to the next.

Test 1: Hypothesis Test

1. H_0: P = .308

2. H_A: P > .308

Teaching note: I use P to represent Jake's shooting %, which is a proportion, or P. The null hypothesis is no change. For Jake, no change means he is still shooting 30.8%, or P = .308. For the alternative hypothesis, I look back at the original question which asked if Jake is statistically **better** and that leads me to choose the >

3. **Test statistic and its value:** z = 2.02

4. *p*-value: .022

5. **Statistical conclusion:** Since the *p*-value < .05, I reject H_0

6. **Interpretation:** Jake's 3-point shooting is statistically better, the differences we see are not random variation.

Test 2: Confidence Interval

7. **95% Confidence interval:** (.30672, .44597)

8. **Statistical conclusion:** I am 95% confident Jake will make between 31% and 45% of his 3-point shots.

9. **Interpretation:** .308 is in the confidence interval, Jake could still be a 30.8% shooter. Jake is NOT statistically better this year.

   Teaching note: Always use the old number to see if it is in the interval or not. The same number as we used in our H_o. I look to see if .308 is in the confidence interval. Why? Because I am 95% confident Jake will be in this interval, if the old number is in the interval, Jake could still be the old number. He could still be a 30.8% shooter. He is not statistically better. Every problem we do, if they are in the confidence interval, they have not changed.

   Now, you have done 2 tests on Jake to determine if he is statistically better. Confirm means both tests give you the same result. To answer this always look at your two interpretations. Those are #6 and #9 above. Take a second and look up at 6 and 9 ....

Confirmation

10. **Do the results of the two tests confirm each other?** No, the hypothesis test shows Jake is statistically better, and the confidence interval shows he is NOT statistically better. The two tests give us different results.

   Teaching note: While the two tests usually do confirm each other (they are both statistical tests, and both testing the same data), sometimes they give us different results. Here, the reason was the confidence interval. Jake was ALMOST out of the interval, he missed being out by .00128! IF he had been out of the interval, then we would have said he had improved! And therefore confirming the hypothesis test.

Answer without teaching notes

The above answer includes my teaching notes. Here is what your answer using the **10-step answer format** should look like:

Test 1: Hypothesis Test

1. H_0: P = .308
2. H_A: P > .308
3. **Test statistic and its value:** z = 2.02
4. ***p*-value:** .022
5. **Statistical conclusion:** Since the *p*-value < .05, I reject H_0
6. **Interpretation:** Jake's 3-point shooting is statistically better, the differences we see are not random variation.

Test 2: Confidence Interval

7. **95% Confidence interval:** (.30672, .44597)
8. **Statistical conclusion:** I am 95% confident Jake will make between 31% and 45% of his 3-point shots.
9. **Interpretation:** .308 is in the confidence interval, Jake could still be a 30.8% shooter. Jake is NOT statistically better this year.

Confirmation

10. **Do the results of the two tests confirm each other?** No, the hypothesis test shows Jake is statistically better, and the confidence interval shows he is NOT statistically better. The two tests give us different results.

OPTIONAL: CONFIDENCE INTERVALS BY HAND WITH THEO EPSTEIN

Optional section where Theo Epstein delves deeper into 1-proportion statistical tests done by hand.

David Ortiz came to the Red Sox via a trade with the Minnesota Twins after the 2002 season. He was an average hitter with the Twins, hitting .266 for six full seasons in Minnesota. He was not expected to do much when newly-hired GM Theo Epstein traded for

him.

Instead, Ortiz played for 14 years with the Red Sox, winning seven silver slugger awards, was in the top five of the MVP voting five times, was a 10-time all-star, and won three World-Series rings with the team. He then retired a Red Sox after his 20-year career. Those stats do not reflect his presence in the city of Boston, where he will always be remembered for his 2013 marathon bombing "This is our f***ing city" speech.

David's second-most trending speech was his 2020 guest-lecturer speech at Springfield College. His topic was "How to construct a confidence interval by hand."

According to David, a confidence interval is formed by taking a point estimate (like David Ortiz's sample batting average of .266 with the Twins) and adding and subtracting a "margin of error." So .266 +/- the margin of error. We have the point estimate (.266), now we need to calculate the margin of error.

$$\text{The margin of error} = z^*\sqrt{\frac{\hat{p}\left(1-\hat{p}\right)}{n}}$$

where $\hat{p}$ is the sample proportion (p-hat) and n is the sample size.

First we need z*. The z* is the critical z value. We get that value from a z table. This is the same z you used as the test statistic for 1-proportion testing. I went to the z table, and went to the column for 95% confidence, because we will always use 95% in our class for confidence intervals. The critical z value from the table was 1.96.

P-hat is our sample proportion. Theo had David's proportion of hits with the Twins (the sample). In his six seasons with Minnesota, Ortiz had 393 hits in 1477 at bats, or 393/1477, or .266. P-hat = .266

1 – p-hat is easy. It is 1 - .266, or .734.

The n is David's sample size. The number of at-bats David had in Minnesota was 1477. So n = 1477.

Now, Theo (and his minions) did the math.

$$\textbf{Margin of error} = 1.96 \times \sqrt{\frac{.266 \times .734}{1477}}$$

$$\textbf{Multiply:} \quad = 1.96 \times \sqrt{\frac{.1952814}{1477}}$$

$$\textbf{Divide:} \quad = 1.96 \times \sqrt{.00013221488}$$

Square root $= 1.96 \times .0114984731$

Multiply: $= .0225$

Now he took David's .266 with the Twins. He added and subtracted the margin of error. .266 + .0225 = .2885 = .289. The .289 is the high end of what Theo expected for David.

.266 - .0225 = .2435 = .244. This is the low end of Theo's expectations. The 95% confidence interval is (.244, .289).

Theo was 95% confident when he traded for Ortiz that David would hit for between .244 and .289 with the Red Sox.

Note: In his 14 years with the Sox, David had 2079 hits in 7163 at-bats, or 2079/7163 = .290. David outperformed Theo's expectations as he was outside (and above) the confidence interval.

Let's compare our manual confidence interval to Minitab. For Minitab, I chose Stat, Basic Statistics, 1 Proportion, entered David's proportion with the Twins (393 out of 1477), hit OK, and the results were:

Test and CI for One Pr...   ∨ ✕

⊞ WORKSHEET 1

Test and CI for One Proportion

Method

p: event proportion
Exact method is used for this analysis.

Descriptive Statistics

N	Event	Sample p	95% CI for p
1477	393	0.266080	(0.243688, 0.289406)

We had (.244, .289) for our manual step-by-step 95% confidence interval and Minitab rounds off to the same (.244, .289) for their 95% confidence interval. You can now see where the confidence interval comes from and how it is calculated, but we can just take advantage of Minitab's calculator, it is much easier and quicker and more precise!

1-sample mean statistical test

Say you were standing with one foot in the oven and one foot in an ice bucket. According to the percentage people, you should be perfectly comfortable.

— *Bobby Bragan, baseball player, coach, manager*

PREVIEW: ANSWER "WILL THEY BE STATISTICALLY BETTER?" WITH ONE CHUNK OF DATA

We will do a Springfield College women's lacrosse example together, then I will give you three questions to answer on your own. Then, in case you need a little more repetition, I will give you a 1-sample mean bonus problem. All five of these questions are exactly like the 1-sample mean question on Exam 3. Just a different athlete or team. You will use the same **10-step answer format** you learned in Chapter 20 to answer them, and every question through Exam 3.

GUIDED EXAMPLE FOR 1-SAMPLE MEAN WITH SC WOMEN'S LAX

One of Springfield College's strongest teams this spring (2020) was women's lacrosse. Last year they were 13-7, including a 10-game winning streak. After a semifinal OT win at MIT, they beat regular-season champs Babson in Wellesley for the NEWMAC championship! They then played in the NCAA tournament, finally ending their season with a loss to Franklin & Marshall in a second-round contest. The team was poised to build on that in 2020 before the pandemic.

Those were the fun facts. Here is the question I will use in the guided example to show you how to find the test statistic, *p*-value and 95% confidence interval in Minitab.

Q The 2020 season for the Springfield College women's lacrosse team ended after five games. We will investigate the team's goals per game. Last year's team was a rocket ship, scoring an average of 12.7 goals per game. So hard to match. In the five games in 2020, the team scored 18, 11, 12, 1, and 4 goals. The question for us is, Will this year's team score statistically **fewer** goals?

In Chapter 21, we were able to do 1-proportion tests in Minitab. We had Tom Brady's 373 completions out of 613 attempts. That was a proportion because we could write it as a fraction (373/613), a decimal (.60848), or a percent (60.85%). In this question we do not have any fractions, decimals or percents. We cannot do a proportion test.

What do we have? A list of numbers! We have the goals scored for the 2020 Springfield College women's lax team. Going back to mini-putt which we did in day two of the semester, the best way to summarize numbers in stats is with an average, aka a mean. So no proportion, but a mean.

We also only have 1 sample (Springfield College women's lax). Ohhhh, the title of the chapter is 1-Sample Mean Statistical Tests! We are there.

The last question before we start is which test statistic do we use for 1-sample tests? I would love to use z again. It worked well with proportions. But I cannot use z with means. The reason is I don't have the standard deviation for the population. I need a different test statistic. In statistics we have another distribution that does not need the population standard deviation, and almost matches z's normal approximation. In statistical terms, it's only off by one degree of freedom. Excellent, that works for us. Therefore, the test statistic will be t.

Minitab step-by-step for 1-sample mean

In Minitab we have two options for 1-sample mean tests, z or t. We just used z for proportions, and would love to use z for every test. But as I just mentioned, we have to use t instead. In Minitab therefore, we will use the 1-Sample t option. Here are the choices on Minitab:

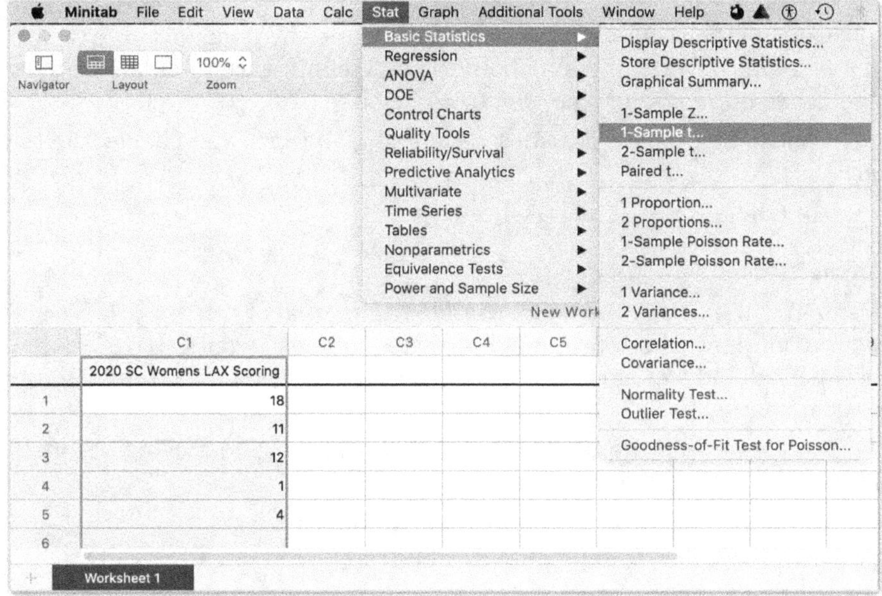

Enter data in Minitab

Label C1 2020 SC Women's Lax Scoring. The question gives us the goal from their five games, 8, 11, 12, 1, and 4. Enter those under C1.

Choose Minitab options

Now we can find our test statistic, p-value and 95% confidence interval we need for this problem.

Choose Stat from the menu bar then Basic Statistics and 1-Sample t....

That brings up our One-Sample t for the Mean screen. It defaults to One or more samples, each in a column. Click in the box below it, and that will bring up C1 in the list on the left. Double click on C1 to move it over.

Check Perform hypothesis test. This gives us access to the box next to Hypothesized mean where we'll put our old number. The number you will put with the H_0 in a second. Which I will give you in the question. In this case, last year (old number) the women scored 12.7 goals per game, so the 12.7 is our Hypothesized mean.

Then click options. Our One-Sample t: Options screen is short. It is even shorter because it defaults to a confidence level of 95.0, which is what we want to do! We do change the Alternative hypothesis, though.

Our problem asked if the women were scoring **fewer goals**. Fewer mean less. Fewer means <. Change ≠ to the top choice, less than, <.

Then hit OK, OK.

On your worksheet are the test statistic and the *p*-value. t = -1.16 and the *p*-value = .156. But what about the 95% confidence interval? Ohhhh, go back to options, and change < to ≠. Just like this:

Choose Stat, Basic Statistics, 1-sample t, Options, and change the H_A to ≠

Hit OK, OK.

Done!

Minitab will give you your 95% confidence interval of (.81, 17.59). We are 95% confident the Springfield College women would have scored between .8 and 17.6 goals per game, if they had been able to play a full season in 2020.

Notice we read the confidence interval just as is, no moving decimals like we did in proportions.

Here is a combined picture of your Minitab screens:

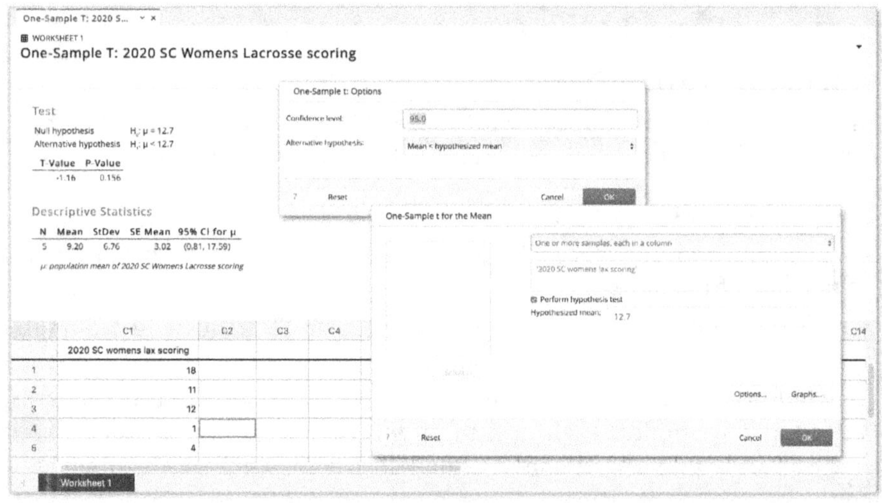

Answer to guided example

Minitab calculated the test statistic (t = -1.16), the *p*-value (.156) and the 95% confidence interval (.81, 17.59) for us. Now we are ready to answer our guided-example question.

Here's the question again: The 2020 season for the Springfield College women's lacrosse team ended after five games. We will investigate the team's goals per game. Last year's team was a rocket ship, scoring an **average of 12.7** goals per game. So hard to match. In the five games in 2020, the team scored 18, 11, 12, 1, and 4 goals. The question for us is, Perform a hypothesis test and a confidence interval to determine if this year's team will score statistically **fewer** goals?

The question asks us to perform two separate tests on the same data. If they give us the same result, the tests confirm each other. But different results mean they do not confirm each other.

Using the Minitab output we just found, here is our guided example answer with my notes.

Our first test is the hypothesis test.

Test 1: Hypothesis Test

1. H_o: M = 12.7
   I cannot use P because there is no proportion in the problem. But I do see an average (bolded in the question above), for us that is fine! An average we called a mean. Therefore the M stands for the mean. The old number for women's lax was 12.7 goals per game. M = 12.7 is our way of saying, don't worry, the team will still score 12.7 goals per game this season!

2. H_A: M < 12.7
   Keep the M, keep the 12.7, just change the sign based on my question. My question says **fewer goals**, so we choose <. FYI, when I write the question, I look up the numbers. In this case the team had scored 9.2 goals per game so far this year, hence the "fewer goals" in the question. I do the same for all the exams.

3. **Test statistic and its value:** t = -1.16
   A negative test statistic is OK, it means the women are scoring fewer goals. But we are testing to see if they are **statistically** lower in scoring. t is our new test statistic though, and it comes from the Normal distribution, but using n-1. I cannot use z because I do not have the population standard deviation.

4. *p*-value: .156
   There is a 15.6% chance the women's lacrosse team will still score 12.7 goals per game in 2020, given the sample data.

5. **Statistical conclusion:** Since the *p*-value > .05, I fail to reject H_o. No changes to our statistical conclusion with 1-sample mean tests.

6. **Interpretation:** The Springfield College women's lax team did not statistically score fewer goals in 2020. The difference we see is just

random variation.

They still could have scored 12.7 goals per game if the season had continued. There was a difference, last year they scored 12.7 goals per game, this year 9.2. But, based on the hypothesis test, it was not a statistical difference, it was just random variation!

Our second test is the 95% confidence interval.

Test 2: Confidence Interval

7. **95% Confidence interval:** (.80613119, 17.593869)
No change. Just record the low and high ends of the interval here. in Step 7.

8. **Statistical conclusion:** I am 95% confident that SC women's lax would have scored between .8 and 17.6 goals per game in 2020.

   Much easier than proportions, where we had to move the decimal point two places to change it to a percent. With means we do not move the decimal point. We keep the number just as is.

9. **Interpretation:** Since 12.7 is in the interval, the women's lacrosse team is not scoring statistically fewer goals.

   Just like in 1-proportion testing we look to see if the null hypothesis (the old number) is in the interval. If it is in the confidence interval, there is no statistical difference. Here it means the team could still score 12.7 goals per game. But, if it is outside the interval, then they are statistically different.

   We have done two tests on the same data. Did we get the same result? In step 10 we check to see if we did get the same result. Do the two tests confirm each other?

Confirmation

10. **Do the results of the two tests confirm each other?** Yes, the confidence interval and the hypothesis test both show the women's lax team was not scoring statistically fewer goals in 2020.

    They confirm because parts 6 and 9 both give the same answer, no statistical difference. They are confirming that there is no statistical difference between this year's team and last year's team.

Answer without teaching notes

The above example for Springfield College women's lax has all my notes. All you need to print out for the next three questions is the **10-step answer format**, like this:

Test 1: Hypothesis Test

1. H_0: M = 12.7

2. H_A: M < 12.7

3. **Test statistic and its value:** t = -1.16

4. *p*-value: .156

5. **Statistical conclusion:** Since the *p*-value > .05, I fail to reject H_0.

6. **Interpretation:** The Springfield College women's lax team did not statistically score fewer goals in 2020. The difference we see is just random variation.

Test 2: Confidence Interval

7. **95% Confidence interval:** (.80613119, 17.593869)

8. **Statistical conclusion:** I am 95% confident that SC women's lax would have scored between .8 and 17.6 goals per game in 2020.

9. **Interpretation:** Since 12.7 is in the interval, the women's lacrosse team is not scoring statistically fewer goals.

Confirmation

10. **Do the results of the two tests confirm each other?** Yes, the confidence interval and the hypothesis test both show the women's lax team was not scoring statistically fewer goals in 2020.

PRACTICE PROBLEMS

Please answer the three questions below (after doing the guided example on women's lax) using the **10-step answer format** from Chapter 20.

SC men's lacrosse also went to the NCAA Division III tournament in 2019. And 2018. And 2017.... And 2008. Yes, 12 straight years of making NCAAs! We will look at men's lacrosse scoring (goals per game). In 2019, the men scored 11.81 goals per game. In the four games they played in 2020, they scored 10, 8, 6 and 9 goals. Perform a hypothesis test and a 95% confidence interval to ask, Are the men scoring statistically less in 2020?

SC women's basketball had a good year, making it to the semifinals of the NEWMAC tournament, before losing a close game to Smith College. They were led by Northeast second-teamer Alex Goslin. We will look at the team's rebounding. In 2018-19 they averaged 39.9 rebounds per game. I looked at the rebounds per game for this year (listed below), and they looked better. Perform a hypothesis test and a 95% confidence interval to ask, Are the women statistically better rebounders this year?

2020 SC Women's Hoops rebounds:

52	45	42	36	51	41	37	42	38
46	53	53	47	36	31	48	34	40
44	48	34	41	44	37	44	41	42

SC men's basketball had a great four-year run, led by senior Jake Ross. At one point this year he led Division III in scoring, and had the record setting 55-point game here against Coast Guard. Last year Jake scored 23.8 points per game. Perform a hypothesis test and a 95% confidence interval to ask, Has Jake's scoring statistically increased in 2020?

2020 Jake Ross scoring by game:

34	41	33	31	32	19	26	31	30
20	10	24	16	31	14	9	30	20
13	55	29	14	10	32	35	44	16
29								

Answers to practice problems

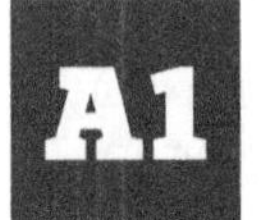

SC men's lacrosse also went to the NCAA Division III tournament in 2019. And 2018. And 2017.... And 2008. Yes, 12 straight years of making NCAAs! We will look at men's lacrosse scoring (goals per game). In 2019, the men scored 11.81 goals per game. In the four games they played in 2020, they scored 10, 8, 6 and 9 goals. Perform a hypothesis test and a 95% confidence interval to ask, Are the men scoring statistically less in 2020?

Test 1: Hypothesis Test

One-Sample T: SC Men...   ⌄ ✕

⊞ WORKSHEET 5

One-Sample T: SC Mens Lax

Test

Null hypothesis	$H_0: \mu = 11.81$
Alternative hypothesis	$H_1: \mu < 11.81$

T-Value	P-Value
-4.17	0.013

1. **H_0:** M = 11.81
2. **H_A:** M < 11.81
3. **Test statistic and its value:** t = -4.17
4. **_p_-value:** .013
5. **Statistical conclusion:** Since the _p_-value < .05, I reject H_0
6. **Interpretation:** The men are scoring statistically less in 2020.

Test 2: Confidence Interval

One-Sample T: SC Men...   ⌄ ✕

⊞ WORKSHEET 5

One-Sample T: SC Mens Lax

Descriptive Statistics

N	Mean	StDev	SE Mean	95% CI for μ
4	8.250	1.708	0.854	(5.532, 10.968)

μ: population mean of SC Mens Lax

7. **95% Confidence interval:** (5.532, 10.968)
8. **Statistical conclusion:** I am 95% confident the men's lacrosse team is scoring between 5.53 and 10.97 goals per game in 2020.

9. **Interpretation:** Since 11.81 is not in the interval, and the interval is less than 11.81, the men are scoring statistically less in 2020.

Confirmation

10. **Do the results of the two tests confirm each other?** Yes, both the confidence interval and the hypothesis test show the men's lacrosse team is scoring statistically less in 2020.

SC women's basketball had a good year, making it to the semifinals of the NEWMAC tournament, before losing a close game to Smith College. They were led by Northeast second-teamer Alex Goslin. We will look at the team's rebounding. In 2018-19 they averaged 39.9 rebounds per game. I looked at the rebounds per game for this year (listed below), and they looked better. Perform a hypothesis test and a 95% confidence interval to ask, Are the women statistically better rebounders this year?

2020 SC Women's Hoops rebounds:

52	45	42	36	51	41	37	42	38
46	53	53	47	36	31	48	34	40
44	48	34	41	44	37	44	41	42

Test 1: Hypothesis Test

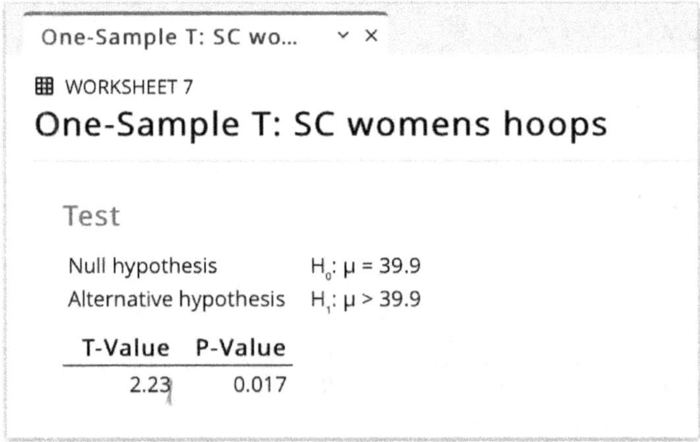

One-Sample T: SC wo…

WORKSHEET 7

One-Sample T: SC womens hoops

Test

Null hypothesis	$H_0: \mu = 39.9$
Alternative hypothesis	$H_1: \mu > 39.9$

T-Value	P-Value
2.23	0.017

1. **H₀:** $M = 39.9$

2. **Hₐ:** $M > 39.9$

3. **Test statistic and its value:** $t = 2.23$

4. ***p*-value:** .017

5. **Statistical conclusion:** Since the *p*-value $< .05$, I reject H_0

6. **Interpretation:** The women are statistically better rebounders in 2020.

Test 2: Confidence Interval

One-Sample T: SC wo...   ⌄ ✕

▦ WORKSHEET 7

One-Sample T: SC womens hoops

Descriptive Statistics

N	Mean	StDev	SE Mean	95% CI for μ
27	42.48	6.03	1.16	(40.10, 44.87)

μ: population mean of SC womens hoops

7. **95% Confidence interval:** (40.10, 44.87)

8. **Statistical conclusion:** I am 95% confident the women will get between 40.1 and 44.9 rebounds per game in 2020.

9. **Interpretation:** Since 39.9 is NOT in the interval, and the entire interval is higher, the women are statistically better rebounders in 2020.

Confirmation

10. **Do the results of the two tests confirm each other?** Yes, both the confidence interval and the hypothesis test show that the women are statistically better rebounders in 2020.

SC men's basketball had a great four-year run, led by senior Jake Ross. At one point this year he led Division III in scoring, and had the record setting 55-point game here against Coast Guard. Last year Jake scored 23.8 points per game. Perform a hypothesis test and a 95% confidence interval to ask, Has Jake's scoring statistically increased in 2020?

2020 Jake Ross scoring by game:

34	41	33	31	32	19	26	31	30
20	10	24	16	31	14	9	30	20
13	55	29	14	10	32	35	44	16
29								

Test 1: Hypothesis Test

One-Sample T: SC Men…   ⌄ ✕

⊞ WORKSHEET 8

One-Sample T: SC Mens Hoops

Test

Null hypothesis	H_0: $\mu = 23.8$
Alternative hypothesis	H_1: $\mu > 23.8$

T-Value	P-Value
1.04	0.153

1. **H_0:** M = 23.8
2. **H_A:** M > 23.8
3. **Test statistic and its value:** t = 1.04
4. **_p_-value:** .153
5. **Statistical conclusion:** Since the _p_-value > .05, I fail to reject H_0
6. **Interpretation:** Jake's scoring has not statistically increased in 2020. The differences we see are just random variation.

Test 2: Confidence Interval

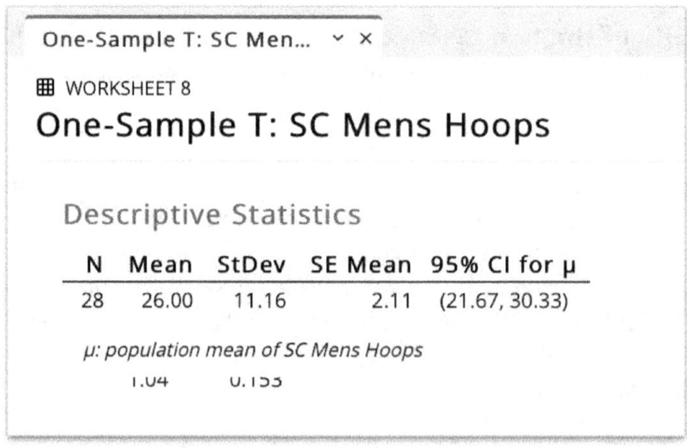

One-Sample T: SC Men…   ⌄ ✕

⊞ WORKSHEET 8

One-Sample T: SC Mens Hoops

Descriptive Statistics

N	Mean	StDev	SE Mean	95% CI for μ
28	26.00	11.16	2.11	(21.67, 30.33)

μ: population mean of SC Mens Hoops

1.04   0.153

7. **95% Confidence interval:** (21.67, 30.33)

8. **Statistical conclusion:** I am 95% confident Jake will score between 21.7 and 30.3 points per game in 2020.

9. **Interpretation:** Since 23.8 is in the interval, Jake's scoring has not statistically increased.

Confirmation

10. **Do the results of the two tests confirm each other?** Yes, both the confidence interval and the hypothesis test show that Jake's scoring has not statistically increased in 2020.

BONUS PROBLEM WITH BROOKS ORPIK

Q+ Brooks Orpik is one of the best US-born (San Fransisco) defensemen to have played in the NHL. Brooks played three years at Boston College before being drafted by the Pittsburgh Penguins. He then played 11 seasons with the Penguins before moving on to the Washington Capitals. He finished his career in 2019 after playing five seasons in our nation's capital. He finished with 156 career playoff games and 2 Stanley Cups.

Brooks was not Tory Krug, but he did accumulate 12 points per season with the Penguins. In his five seasons with the Capitals, he scored: 19, 10, 14, 10, and 9 points. Was Brooks a statistically **better** offensive player with the Capitals?

Answer to bonus problem

Test 1: Hypothesis Test

1. H_0: M = 12.0

2. H_A: M > 12.0

3. **Test statistic and its value:** t = .215

4. ***p*-value:** .42

5. **Statistical conclusion:** Since *p*-value > .05, I fail to reject H_0

6. **Interpretation:** Brooks was NOT statistically better with the Capitals, the difference we see is just random variation.

Test 2: Confidence Interval

7. **95% Confidence interval:** (7.2355, 17.564)

8. **Statistical conclusion:** I am 95% confident Brooks will score between 7.2 and 17.6 points per season.

9. **Interpretation:** Since 12.0 is in the interval, Brooks is NOT statistically better.

Confirmation

10. **Do the results of the two tests confirm each other?** Yes, both the confidence interval and the hypothesis test show the Brooks was NOT statistically better with the Capitals.

OPTIONAL: CONFIDENCE INTERVALS BY HAND WITH JD

Remember, Part 3 is about inference, using a sample to estimate the population. We have two tests (hypothesis test and confidence intervals) to accomplish that estimate. JD Martinez will now take us on a journey where we can do confidence intervals for means by hand. He will use the confidence interval to apply his career home runs (his sample) to estimate his home runs in 2020.

Here are JD's home run totals for every season of his career:

Year	Home runs	Year	Home runs
2011	6	2016	22
2012	11	2017	45
2013	7	2018	43
2014	23	2019	36
2015	38		

Now that we know what a confidence interval is, what is JD's 95% confidence interval for the home runs he will hit in 2020? Using Minitab, I get a 95% confidence interval for JD of (13.81, 37.53). Here is the output from Minitab:

One-Sample T: JD Mart...   ∨ ✕

⊞ WORKSHEET 1

One-Sample T: JD Martinez career homeruns

Descriptive Statistics

N	Mean	StDev	SE Mean	95% CI for μ
9	25.67	15.43	5.14	(13.81, 37.53)

μ: population mean of JD Martinez career homeruns

	C1	C2	C3	C4	C5
	Year	JD Martinez homeruns			
1	2011	6			
2	2012	11			
3	2013	7			
4	2014	23			
5	2015	38			

The (13.81, 37.53) means I am 95% confident JD will hit between 13.81 and 37.53 home runs in 2020.

But, how do we calculate JD's confidence interval by hand? Just like proportions, we generally take a central point, then add and subtract (+/-) a margin of error. Proportions had a formula David Ortiz could use to build his confidence interval by hand. Means also need a formula JD can use.

The confidence interval formula for means is different, it is:

$$\bar{X} \pm t\frac{s}{\sqrt{n}}$$

Where X-bar is the sample mean, t is the critical value from the t distribution, s is the sample standard deviation, and n is the sample size.

JD told us to find his confidence interval. We should take a central point, and +/- a margin of error. Ohhhhhh, the second part of the formula is the margin of error.

To construct the confidence interval by hand therefore, we need X-bar, t, s and n.

X-bar is the central point. It is the sample mean. To get JD's sample mean I add up the home runs he has hit over his career, then divide by 9, the number of seasons he has played.

$$\bar{X} = \frac{\text{JD's homeruns}}{\text{number of seasons}} = \frac{231}{9} = 25.67$$

t is the critical value from the t distribution. JD goes to the t distribution tables in his stats book, and looks up his t value. Using his 95% confidence level, and degrees of freedom of 8 (n − 1), JD finds his t value of 2.306.

t = 2.306.

s is the sample standard deviation. He loads his nine years' worth of data into the calculator on this phone, and calculates his sample's standard deviation. He gets 15.427.

s = 15.427.

n is the sample size. JD has nine years' worth of data, so his sample size is 9.

n = 9.

Now he can fill in his values:

X-bar (the sample mean) = 25.67

t (the critical value) = 2.306

s (the standard deviation of our sample) = 15.427

n (the sample size) = 9

into the formula: $\bar{X} \pm t \dfrac{s}{\sqrt{n}}$

Plug in: $= 25.67 \pm 2.306 \times \dfrac{15.427}{\sqrt{9}}$

Square root: $= 25.67 \pm 2.306 \times \dfrac{15.427}{3}$

Divide: $= 25.67 \pm 2.306 \times \mathbf{5.142}$

Multiply: $= 25.67 \pm \mathbf{11.86}$

Subtract: $= \mathbf{13.81}$

And **add**: $= \mathbf{37.53}$

The low end of JD's confidence interval is 13.81. The high end is 37.53. Therefore, his 95% confidence interval by hand is (13.81, 37.53). He is 95% confident he will hit between 13.81 and 37.53 home runs in 2020.

Note the (13.81, 37.53) JD calculated by hand agrees with Minitab's (13.81, 37.53). Minitab was easier, quicker, and more precise. We will stick with Minitab.

Now we should feel pretty comfortable with 1-sample mean problems. Let's try 2-sample means!

JD Fun Fact

JD's career high was 45. Do I think he will hit 45 again in 2020? No. I am 95% confident he will hit between 13.8 and 37.5 home runs in 2020. 45 is not in the confidence interval, so using statistical inference I decide that he will not hit 45 home runs in 2020.

Last year he hit 36. Do I think he could hit 36 home runs again in 2020? Yes! Again, I am 95% confident he will hit between 13.8 and 37.5 home runs in 2020. The interval contains 36, means he could absolutely hit 36 again!

2-sample means statistical test

I've missed more than 9000 shots in my career. I've lost almost 300 games. 26 times, I've been trusted to take the game winning shot and missed. I've failed over and over and over again in my life. And that is why I succeed.

— Michael "Air" Jordan, basketball legend

PREVIEW: I CAN TELL YOU WHICH TEAM IS BETTER WITH JUST TWO CHUNKS OF DATA

In the last two chapters we did 1-proportion and 1-sample mean problems. Now we are going to do 2-sample means problems, and the next chapter will be 2-proportions problems. In Chapter 23, everything you learned from 1-sample mean testing in Chapter 22 will still apply. We just need to extend our test to cover two samples instead of one.

What was 1-sample mean? When we looked at women's lax in the Chapter 22, it was only one sample. The sample we took from the team was their goals scored. Their sample of goals scored for 2020 was: 18, 11, 12, 1 and 4. Their average goals scored for 2020 was 9.4. Their mean was 9.4. Their 1-sample mean was 9.4.

Now we are going to compare them to another sample. The second sample for teaching purposes will be Springfield College men's lacrosse.

We will use their goals scored in 2020, so we can compare them to the women. The Springfield College men scored 10, 8, 6, and 9 goals in 2020. That is my second sample. The Springfield College women provided the first sample, the men will provide the second sample. Hence, a 2-sample means problem.

Test 1: Hypothesis test

The first difference in the hypothesis test is I no longer compare my 1-sample to an old number. I compare the two samples I have directly. In this example, one sample is Springfield College women's lax 2020 scoring. The second sample is the men's lax 2020 scoring. I am going to compare the two samples to see if there is a statistical difference between the two teams.

When you look at the H_0 in the answer format sheet coming up, you will notice I no longer compare the mean (M) to an old number. I compare it to the mean of the second team. In general, I will teach it as $M_1 = M_2$, but when I do the specific example, we will talk about that team. For the assignment, you would write: $M_{wlax} = M_{mlax}$ (which is stats for saying the mean scoring for the women is the same as mean scoring for the men).

On your questions, I will look up the data. For instance the women are scoring 9.4 goals per game in 2020, the men 8.25, so I will phrase the question, Are the 2020 Springfield College women's lax team statistically scoring **more** than the men's team? You do NOT have to look those numbers up. I do when writing the question. But now you know to phrase the H_A: $M_{wlax} > M_{mlax}$ because I wrote in the question, Are the women scoring **more**. More means >.

When doing the hypothesis test, we need a test statistic. With 1-sample mean, we learned about t. We used t for our test statistic because of the mean. Now we have two means, do we get to keep t? Yes! Whenever we are working with means, we will use t. The test statistic of t stays.

That summarizes the changes and similarities (we keep t as our test statistic) for hypothesis testing. How about the differences in the 95% confidence interval between 1-sample and 2-sample means?

Test 2: Confidence interval

There are 1-sample and 2-sample confidence interval differences. With one sample, we looked to see if the old number was in the confidence interval. Not with two samples. Anytime you have two samples you will use the sign rule.

In general, the sign rule is:

If the confidence interval goes from a negative number to another negative number, like (-40, -2), there is a difference, the first mean is lower.

If the confidence interval goes from positive to positive, like (2, 40), there is a difference, the first number is higher.

If the confidence interval goes from negative to positive, like (-40, 2), there is NO difference. This is the same as thinking, if the interval contained zero, there was NO difference.

Sign rule summary:

- Negative to negative, there is a difference.
- Positive to positive, there is a difference.

If your signs are the same, there is a difference. What is the difference though?

- Negative to negative means the **first sample** is less, or lower.
- Positive to positive means the **first sample** is more, or higher.

Notice that whichever sample you put in **first** is the one you will talk about **first**.

- And negative to positive, there is NO difference.

I will use a Red Sox vs. Yankees example to illustrate the wording for confidence intervals.

Sign-rule specific example

If we did a confidence interval for the Red Sox runs scored in 2019 compared to the Yankees runs scored in 2019 that is a 2-sample problem. Remember to always talk about what you put in first, and I put the Red Sox in first. Here are the three possible answers using the numbers above I created for teaching purposes. To make it easier to picture, first we will use integers. Then we'll use the real life numbers, which are, of course, decimals! Finally, notice I am using the same 7, 8 and 9 from our **10-step answer format** we learned in Chapter 20.

First, the integers. If the Minitab confidence interval was (-40, -2), then my answers would have been:

7. **95% Confidence interval:** (-40, -2)

8. **Statistical conclusion:** I am 95% confident the Red Sox scored between 2 and 40 fewer runs (note: I used fewer because of the negative signs) than the Yankees last year.

9. **Interpretation:** Since the confidence interval goes from negative to negative, there is a difference, the Red Sox are scoring fewer runs.

If the confidence interval had been (-40, 2), then my answers would have been:

7. **95% Confidence interval:** (-40, 2)

8. **Statistical conclusion:** I am 95% confident the Red Sox will score between 40 runs fewer (because of the negative) and 2 runs more (because of the positive) than the Yankees.

9. **Interpretation:** Since the confidence interval goes from negative to positive (because of that, notice the confidence interval contains 0) there is NO difference in runs scored for the Red Sox and Yankees.

If the confidence interval had been (2, 40), then my answers would have been:

7. **95% Confidence interval:**(2, 40)

8. **Statistical conclusion:** I am 95% confident the Red Sox will score between 2 and 40 runs more (more because both numbers were positive) than the Yankees.

9. **Interpretation:** Since the confidence interval is positive to positive, there is a difference, the Red Sox score more runs than the Yankees.

For the example, I gave you nice easy numbers for teaching purposes. It is much easier to see what's going on with simple numbers. But in real life, the numbers are decimals, and since statistics is about real data, I found the real numbers for you. Well, I let Minitab find the real Red Sox vs. Yankees confidence interval for us.

I loaded the 2019 Red Sox runs scored for all 162 games into C1, and the Yankees into C2. The Minitab picture below shows the first nine games, but there is 162 games worth of data in there. Then I performed the keystrokes you are about to learn in Chapter 23, and we could find the low and high ends of the confidence interval for the 2019 Red Sox vs. Yankees. The Minitab output is here with the 95% confidence interval at the very bottom:

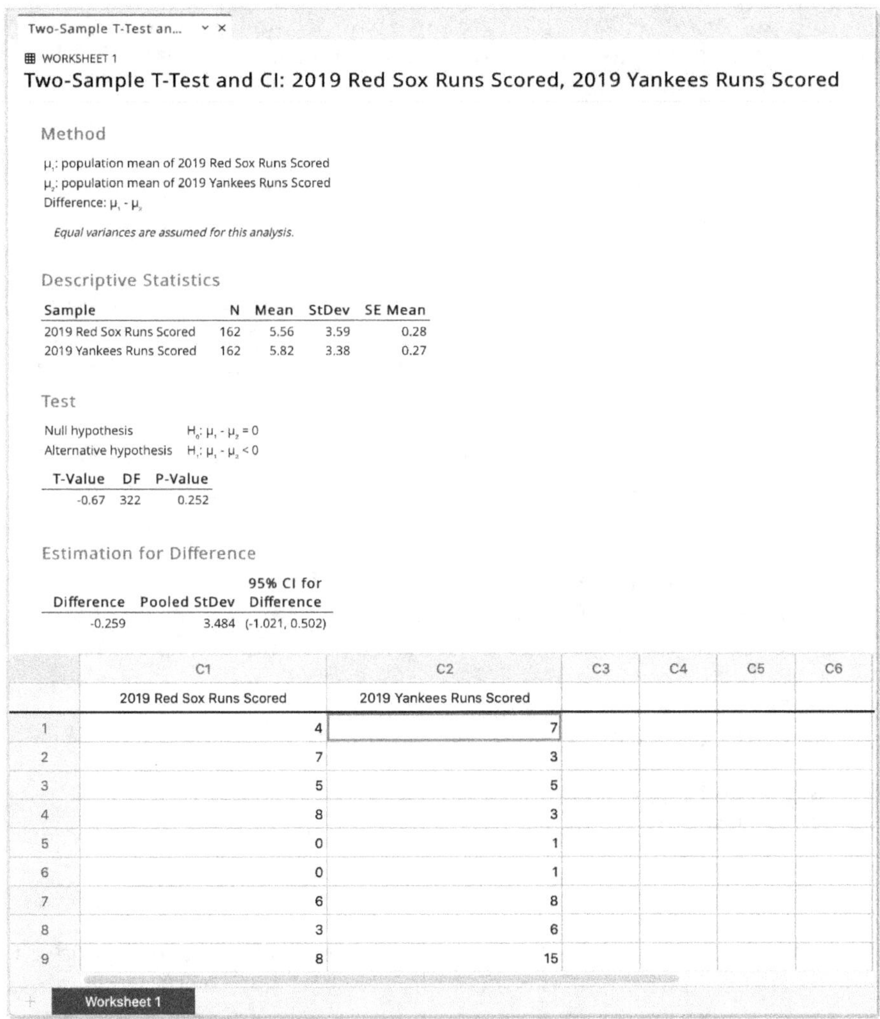

Two-Sample T-Test and CI: 2019 Red Sox Runs Scored, 2019 Yankees Runs Scored

Method

μ_1: population mean of 2019 Red Sox Runs Scored
μ_2: population mean of 2019 Yankees Runs Scored
Difference: $\mu_1 - \mu_2$

Equal variances are assumed for this analysis.

Descriptive Statistics

Sample	N	Mean	StDev	SE Mean
2019 Red Sox Runs Scored	162	5.56	3.59	0.28
2019 Yankees Runs Scored	162	5.82	3.38	0.27

Test

Null hypothesis $H_0: \mu_1 - \mu_2 = 0$
Alternative hypothesis $H_1: \mu_1 - \mu_2 < 0$

T-Value	DF	P-Value
-0.67	322	0.252

Estimation for Difference

Difference	Pooled StDev	95% CI for Difference
-0.259	3.484	(-1.021, 0.502)

	C1	C2	C3	C4	C5	C6
	2019 Red Sox Runs Scored	2019 Yankees Runs Scored				
1	4	7				
2	7	3				
3	5	5				
4	8	3				
5	0	1				
6	0	1				
7	6	8				
8	3	6				
9	8	15				

Check it out Red Sox fans, the low and high ends of the 95% confidence interval for 2019 Red Sox vs. Yankees runs scored was (-1.021, .502), which I answered as follows:

7. **95% Confidence interval:** (-1.021, .502)

8. **Statistical conclusion:** I am 95% confident the Red Sox will score between 1 run less and .5 runs more than the Yankees.

9. **Interpretation:** Since the confidence interval goes from negative to positive, there is no STATISTICAL difference in runs scored between the 2019 Red Sox and Yankees.

I do all ten steps as the bonus question at the end of Chapter 23.

But for now we can put the two tests (hypothesis test and the confidence interval) together and do 2-sample means testing. You have all the tools in your toolbox.

2-sample means answer format

Test 1: Hypothesis Test

1. **H_0:** $M_1 = M_2$
   This is the null hypothesis, where the = sign goes. Because nothing is going on here, the two samples we are investigating are the same, they are =. We say $M_1 = M_2$ where 1 and 2 represent our two samples.

2. **H_A:** $M_1 < M_2$ or $M_1 > M_2$
   This is the alternative hypothesis, where the inequality goes. We will use less than (<) or greater than (>) depending on the wording in the problem. Are they worse? Use <. Have they improved? Use >.

3. **Test statistic and its value:** t =
   We need to use t again, like we did in the 1-sample mean problems. Minitab only calculates t for the 2-sample test, making it easy for us!

4. ***p*-value:**
   Same probability from before that we will compare to .05

5. **Statistical conclusion**
   Either: Since the *p*-value < .05, I reject H_0 or since the *p*-value > .05, I fail to reject H_0

6. **Interpretation**
   If we reject H_0, there is something going on, and we have support for the H_A. H_A is where we put the < or >. Now we have evidence that one teams is different from another, and in which direction. If we fail to reject, nothing is going on, it is just random variation. There is no difference between the two means.

Test 2: Confidence Interval

7. **95% Confidence interval:**
   Just record the low and high ends of the interval here.

8. **Statistical conclusion**
   Talk about M_1 first. I am 95% confident the first mean is less if the confidence was negative to negative, or the first mean is more if the interval is positive to positive. Negative to positive means you are

95% confident the first mean is between some amount less and some amount more. Less and more because of the negative and positive.

9. **Interpretation**

No more old number in the interval. Now it is the rule of signs. If the confidence interval numbers are both negative, you would say, "There is a difference, the first team is less." If the confidence interval numbers are both positive, you would say, "There is a difference, the first team is more." If the confidence interval goes from negative to positive, (or the confidence interval contains 0), you would say, "There is no difference."

Confirmation

10. **Do the results of the two tests confirm each other?**

1-6 above are the old hypothesis test. 7-9 is the new confidence interval test. They are both testing the same data. Do they have the same result? Look at 6 and 9 above. If they have the same result, then confirm each other, and you answer yes. If 6 and 9 give you different answers, then they do not confirm each other, and you answer no.

GUIDED EXAMPLE FOR 2-SAMPLE MEANS WITH SC LACROSSE

This Minitab example will match the guided example coming up. Do this example with me to make sure we get the same answers.

In the 2020 season, the Springfield College women's and man's lax teams scored the number of goals listed below. Perform a hypothesis test and a confidence interval to determine if the women's team was the statistically higher scoring team.

2020 SC Goals

Women's Lax	Men's Lax
18	10
11	8
12	6
1	9
4	

On Minitab, we need to find the test statistic, *p*-value and 95% confidence interval.

We have two samples' worth of means in our guided example. But, they are still means. Just like Chapter 22 (1-sample mean), we will find t. 1-sample or 2-sample, as long as we are doing means, we use t for our test statistic.

Notice in the guided example above we have numbers. We have data. We will put the data in Minitab and use it. Whenever you have data, that is the first step, use it.

Enter data in Minitab

Label C1 SC women's lax scoring and label C2 SC men's lax scoring. If I give you data in the problem, you will need to use it, and now we did!

Choose Minitab options

Choose Stat from the menu bar then Basic Statistics, and scroll down to 2-Sample t.... Notice while you are there, Minitab makes it easy for us, there is only one 2-sample means choice. Click on that.

When you click on it, it defaults to Both samples are in one column. That doesn't fit our problem. We used C1 and C2. Click on the dropdown menu and there is an option for Each sample is in its own column. That's us. Click on that.

Click on the Sample 1 box. Both C1 and C2 will show up. Double click on C1 to move it over to the Sample 1 box. Double-click on C2 to move it to the Sample 2 box. Whichever sample you put in C1 you will talk about first in the confidence interval, so order is important.

Click Options. Confidence level of 95% is OK! Hypothesized difference is 0 is OK! Never change them. But we need to change the Alternative hypothesis to >. (Go back up to our question. I asked if the women were the higher scoring team, higher means greater, higher means >.) Check **Assume equal variances**! *I bold this because on the exam that is the box most students forget to change!* Then hit OK, OK.

Scroll down and you will find the t-value = .27 and right below the *p*-value = .397

Almost done. To get the 95% confidence interval to show up, go back to Stat, Basic Statistics, 2-sample t and click. Go to Options and change the Alternative hypothesis back to ≠ (not equal to). Click OK, OK.

Done!

Scroll down the Minitab output, and your 95% confidence interval will be (-7.35, 9.25).

Now we can do the guided example, using the values we just computed:

t = .27

_p_-value = .397

95% confidence interval = (-7.35, 9.25)

Here is a picture of the Minitab inputs I just walked you through, along with the worksheet with our test statistic, *p*-value and 95% confidence interval outputs:

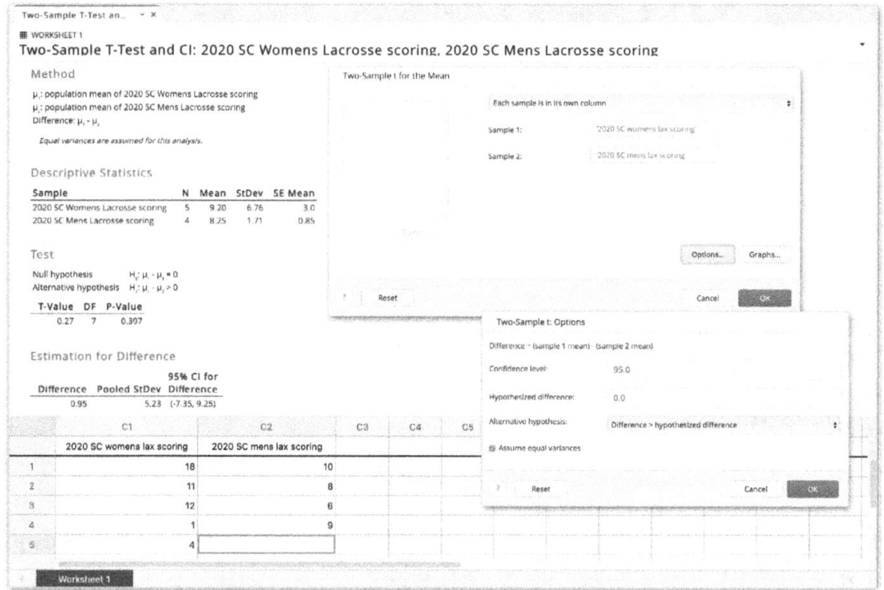

Answer to guided example

I will use the test statistic (t = .27), *p*-value (.397) and 95% confidence interval (-7.35, 9.25) we just calculated on Minitab. I will follow the same **10-step answer format** as I walk you through 2-sample means testing.

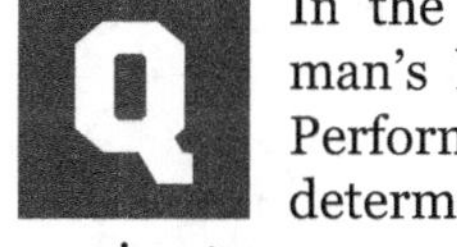

In the 2020 season, the Springfield College women's and man's lax teams scored the number of goals listed below. Perform a hypothesis test and a confidence interval to determine if the women's team was the statistically higher scoring team.

scoring team.

2020 SC Goals

Women's Lax	Men's Lax
18	10
11	8
12	6
1	9
4	

Test 1: Hypothesis Test

1. **H$_0$:** M$_{wlax}$ = M$_{mlax}$
   Note: In a 2-sample means test, we have 2 different samples, each with their own mean. We always set mean 1 = mean 2. Here, we believe the 2 teams are scoring the same number of goals.

2. **H$_A$:** M$_{wlax}$ > M$_{mlax}$
   Note: My question above asks if the women are the higher scoring team. Higher scoring. Greater. >

3. **Test statistic and its value:** t = .27
   Note: I got the t and the .27 from the Minitab step-by-step example.

4. **p-value:** .397

5. **Statistical conclusion:** Since p-value > .05, I fail to reject H$_0$

6. **Interpretation:** The women are NOT a statistically higher scoring team. The differences we see are just random variation.

   Note: Failing to reject means the null (H$_0$) is still there. We could not reject it. H$_0$ said the two teams are equal in terms of scoring, therefore the women are not higher scoring.

Test 2: Confidence Interval

7. **95% Confidence interval:** (-7.35, 9.25)

8. **Statistical conclusion:** I am 95% confident the women will score between 7 goals fewer and 9 goals more than the men.

   Note: The 7 was negative so I used fewer, but the 9 was positive so I more. I rounded but you do not have to round, feel free to keep the numbers exactly.

9. **Interpretation:** Since the confidence interval goes from negative to positive, there is NO statistical difference in scoring between the two teams. The women are NOT a higher scoring team. Remember the sign rule!

Confirmation

10. **Do the results of the two tests confirm each other?** Yes, both the confidence interval and the hypothesis test show there is NO statistical difference in scoring between the two teams. The women were NOT a higher statistically scoring team.

    I answered yes because parts 6 and 9 both show the women are NOT a higher scoring team. The same result means the two tests confirm each other.

PRACTICE PROBLEMS

Use the **10-step answer format** for your answers.

Hint: Keep the two samples in the same order as the problem! Whatever you enter first, that is the one you will talk about first in Step 8 specifically and in the question overall. For instance, in Question 1, you have the Pats OL mentioned first. You load them first in C1, then talk about them first in your answer. In Question 2, you already have the Pats DL entered first, and talk about them first.

In 2019, the New England Patriots offensive and defensive linemen had the following weights. Load the Pats OL into C1 in Minitab and the Pats DL into C2. Then do a hypothesis test and a 95% confidence interval to determine if the Patriots OL weighed statistically **more** than their DL.

Pats OL weights	Pats DL weights
300	250
310	300
335	280
311	315
310	250
310	260
308	305
305	250
310	275

Q2 In 2019, the Springfield College football team had 14 defensive linemen on their roster. Their weights are below. You already have the Pats DL in C2, so you can load the SC DL weights into C3. Then do a hypothesis test and a 95% confidence interval to determine if the Pats DL (C2) weighed statistically **more** the SC DL (C3). Talk about the Pats DL first.

Pats DL weights	SC DL weights
250	235
300	230
280	235
315	250
250	280
260	225
305	210
250	195
275	250
	230
	215
	245
	270
	255

Q3 Lamar Jackson was the NFL MVP in 2019, which he definitely deserved. Enter his regular season passing yards for each game into C1. Then load Tom Brady's passing yards into C2. Then do a hypothesis test and a 95% confidence interval to determine if NFL MVP Lamar Jackson passed for statistically **fewer** yards than our 42-year-old Tom Brady.

Jackson 2019 passing yards		Brady 2019 passing yards	
238	143	221	259
212	236	271	249
145	161	128	334
105	247	169	348
169	267	326	150
222	272	190	306
223	324	216	264
163		285	341

Answers to practice problems

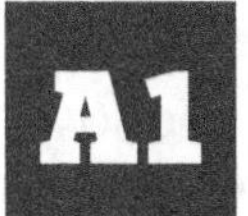

In 2019, the New England Patriots offensive and defensive linemen had the following weights. Load the Pats OL into C1 in Minitab and the Pats DL into C2. Then do a hypothesis test and a 95% confidence interval to determine if the Patriots OL weighed statistically **more** than their DL.

Test 1: Hypothesis Test

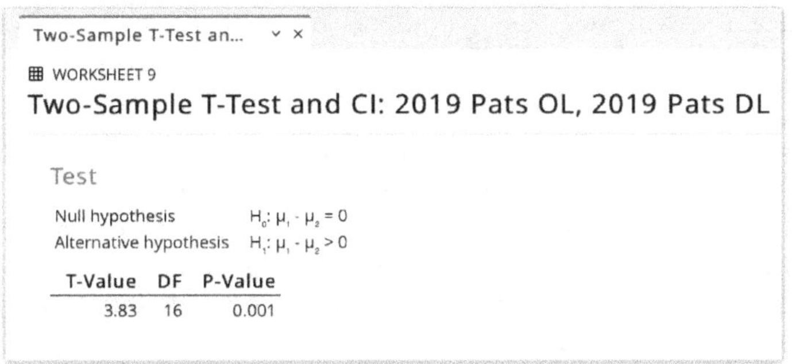

Two-Sample T-Test an... ⌄ ✕

⊞ WORKSHEET 9

Two-Sample T-Test and CI: 2019 Pats OL, 2019 Pats DL

Test

Null hypothesis	$H_0: \mu_1 - \mu_2 = 0$
Alternative hypothesis	$H_1: \mu_1 - \mu_2 > 0$

T-Value	DF	P-Value
3.83	16	0.001

1. **H₀:** $M_{\text{Patriots OL}} = M_{\text{Patriots DL}}$
2. **H_A:** $M_{\text{Patriots OL}} > M_{\text{Patriots DL}}$
3. **Test statistic and its value:** t = 3.83
4. ***p*-value:** .001
5. **Statistical conclusion:** Since the *p*-value < .05, I reject H₀
6. **Interpretation:** The 2019 Patriots OL does weigh statistically more than the Patriots DL. It is NOT random variation.

Test 2: Confidence Interval

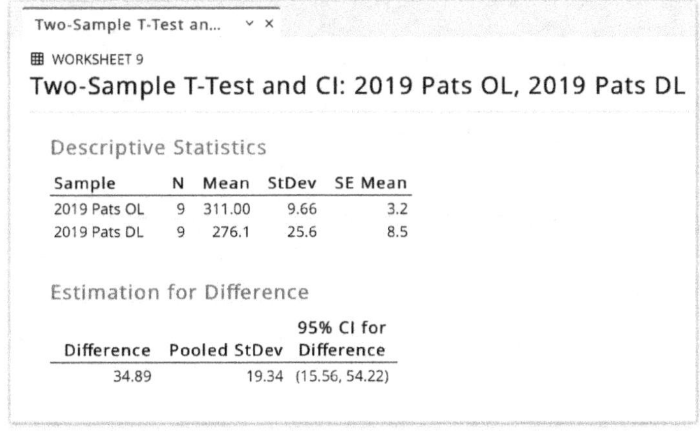

Two-Sample T-Test an... ⌄ ✕

⊞ WORKSHEET 9

Two-Sample T-Test and CI: 2019 Pats OL, 2019 Pats DL

Descriptive Statistics

Sample	N	Mean	StDev	SE Mean
2019 Pats OL	9	311.00	9.66	3.2
2019 Pats DL	9	276.1	25.6	8.5

Estimation for Difference

Difference	Pooled StDev	95% CI for Difference
34.89	19.34	(15.56, 54.22)

7. **95% Confidence interval:** (15.56, 54.22)

8. **Statistical conclusion:** I am 95% confident the 2019 Pats OL weighs between 16 and 54 pounds more than the Pats DL.

9. **Interpretation:** Since the interval goes from positive to positive, there is a statistical difference, the Pats OL does weigh more than the Pats DL.

Confirmation

10. **Do the results of the two tests confirm each other?** Yes, both the confidence interval and the hypothesis test show the Pats OL weighs statistically more than the Pats DL.

In 2019, the Springfield College football team had 14 defensive linemen on their roster. Their weights are below. You already have the Pats DL in C2, so you can load the SC DL weights into C3. Then do a hypothesis test and a 95% confidence interval to determine if the Pats DL (C2) weighed statistically **more** the SC DL (C3). Talk about the Pats DL first.

Test 1: Hypothesis Test

Two-Sample T-Test an…   ∨ ✕

▦ WORKSHEET 10

Two-Sample T-Test and CI: 2019 Pats DL, 2019 SC DL

Test

Null hypothesis	$H_0: \mu_1 - \mu_2 = 0$
Alternative hypothesis	$H_1: \mu_1 - \mu_2 > 0$

T-Value	DF	P-Value
3.76	21	0.001

1. H_0: $M_{\text{Patriots DL}} = M_{\text{Springfield College DL}}$

2. H_A: $M_{\text{Patriots DL}} > M_{\text{Springfield College DL}}$

3. **Test statistic and its value:** $t = 3.76$

4. ***p*-value:** .001

5. **Statistical conclusion:** Since the p-value < .05, I reject H_0

6. **Interpretation:** The 2019 Pats DL does weigh statistically more than the Springfield College DL.

Test 2: Confidence Interval

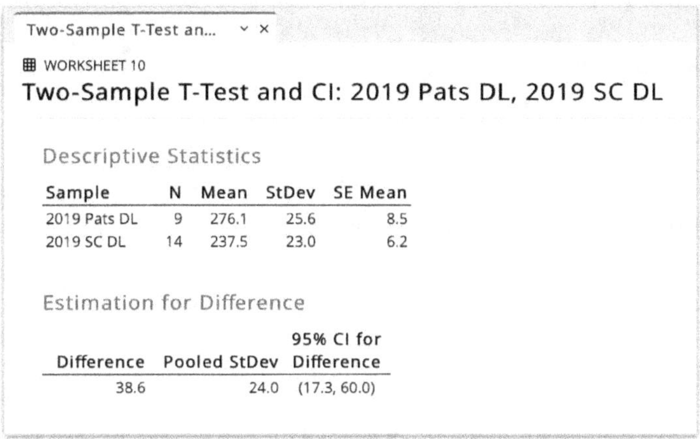

7. **95% Confidence interval:** (7.3, 60.0)

8. **Statistical conclusion:** I am 95% confident the Patriots DL weighs between 17 and 60 pounds more than the Springfield College DL.

9. **Interpretation:** Since the interval goes from positive to positive, there is a difference, the Patriots DL weighs statistically more.

Confirmation

10. **Do the results of the two tests confirm each other?** Yes, both the confidence interval and the hypothesis test show that the 2019 Pats DL weighs statistically more than the Springfield College DL.

Lamar Jackson was the NFL MVP in 2019, which he definitely deserved. Enter his regular season passing yards for each game into C1. Then load Tom Brady's passing yards into C2. Then do a hypothesis test and a 95% confidence interval to determine if NFL MVP Lamar Jackson passed for statistically **fewer** yards than our 42-year-old Tom Brady.

Test 1: Hypothesis Test

1. H_0: MLJ Passing yards = MTB12 Passing Yards
2. H_A: $M_{\text{LJ Passing yards}}$ < $M_{\text{TB12 Passing Yards}}$
3. **Test statistic and its value:** t = -1.94
4. *p*-**value:** .031
5. **Statistical conclusion:** Since the *p*-value < .05, I reject H_0
6. **Interpretation:** Lamar Jackson did pass for statistically fewer yards than TB12.

Test 2: Confidence Interval

Two-Sample T-Test an...   ⌄ ✕

⊞ WORKSHEET 11
Two-Sample T-Test and CI: Lamar Jackson 2019, Tom Brady 2019

Descriptive Statistics

Sample	N	Mean	StDev	SE Mean
Lamar Jackson 2019	15	208.5	59.3	15
Tom Brady 2019	16	253.6	69.6	17

Estimation for Difference

Difference	Pooled StDev	95% CI for Difference
-45.1	64.8	(-92.7, 2.6)

7. **95% Confidence interval**: (-92.7, 2.6)
8. **Statistical conclusion:** I am 95% confident that Lamar Jackson threw for between 93 yards fewer and 3 yards more than Tom Brady.
9. **Interpretation:** Since the interval goes from negative to positive, there is no statistical difference. Lamar Jackson did **NOT** pass for fewer yards than TB12.

Confirmation

10. **Do the results of the two tests confirm each other?** No. The hypothesis test shows Lamar did pass for fewer yards, while the confidence interval shows he did **NOT** pass for fewer yards.

BONUS PROBLEM WITH RED SOX VS. YANKEES

In 2019, the defending World Series champions, the Boston Red Sox, finished 19 games behind the Yankees. While all aspects of the team had issues, we will focus on runs scored. I loaded runs scored for all 162 games for the Red Sox in C1, and for the Yankees in C2. The Red Sox looked like they scored fewer runs. But let's see! Perform a hypothesis test and a 95% confidence interval to determine if the 2019 Red Sox statistically scored fewer runs than the Yankees.

And yes, this is the same question from the beginning of Chapter 23. We used this question to learn about confidence intervals. Which means this is the same Minitab output we saw earlier in the chapter:

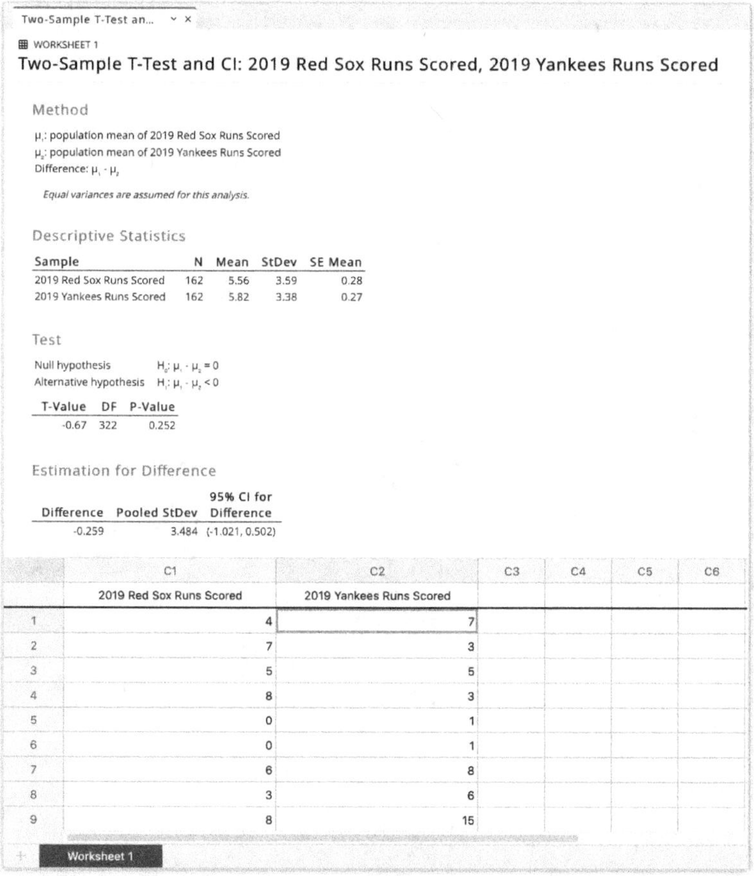

Two-Sample T-Test an...   ⌄ ✕

WORKSHEET 1

Two-Sample T-Test and CI: 2019 Red Sox Runs Scored, 2019 Yankees Runs Scored

Method

μ_1: population mean of 2019 Red Sox Runs Scored
μ_2: population mean of 2019 Yankees Runs Scored
Difference: $\mu_1 - \mu_2$

Equal variances are assumed for this analysis.

Descriptive Statistics

Sample	N	Mean	StDev	SE Mean
2019 Red Sox Runs Scored	162	5.56	3.59	0.28
2019 Yankees Runs Scored	162	5.82	3.38	0.27

Test

Null hypothesis	$H_0: \mu_1 - \mu_2 = 0$
Alternative hypothesis	$H_1: \mu_1 - \mu_2 < 0$

T-Value	DF	P-Value
-0.67	322	0.252

Estimation for Difference

Difference	Pooled StDev	95% CI for Difference
-0.259	3.484	(-1.021, 0.502)

	C1	C2	C3	C4	C5	C6
	2019 Red Sox Runs Scored	2019 Yankees Runs Scored				
1	4	7				
2	7	3				
3	5	5				
4	8	3				
5	0	1				
6	0	1				
7	6	8				
8	3	6				
9	8	15				

Worksheet 1

Answer to bonus problem

Test 1: Hypothesis Test

1. H_0: $M_{RedSox} = M_{Yankees}$

2. H_A: $M_{RedSox} < M_{Yankees}$

3. **Test statistic and its value:** $t = -.67$

4. **p-value:** .252

5. **Statistical conclusion:** Since the p-value > .05, I fail to reject H_0

6. **Interpretation:** The 2019 Red Sox are NOT scoring statistically fewer runs than the Yankees. The difference we see is just random variation.

Test 2: Confidence Interval

7. **95% Confidence interval:** (-1.021, .502)

8. **Statistical conclusion:** I am 95% confident the Red Sox will score between 1 run per game less and .5 runs per game more than the Yankees.

9. **Interpretation:** Since the interval goes from negative to positive, there is NO statistical difference, the 2019 Red Sox are NOT scoring fewer runs than the Yankees.

Confirmation

10. **Do the results of the two tests confirm each other?** Yes, the confidence interval and the hypothesis test both show the 2019 Red Sox are not scoring statistically fewer runs than the Yankees.

OPTIONAL: RED SOX VS. YANKEES CONFIDENCE INTERVAL DETAIL

The formula for the confidence interval for the 2-sample mean is a little busy compared to the formula we used for the 1-sample mean confidence interval (or even 1-proportion testing). That is because we now have two samples to deal with. Here is the formula:

$$\left(\bar{X}_1 - \bar{X}_2\right) \pm t^* \times \sqrt{\frac{s_1^2}{n_1} + \frac{s_2^2}{n_2}}$$

We did the Red Sox first, they are the 1s you see in the formula, and the Yankees were second, they are the 2s. The values I need to plug in are:

X-bar₁ = the 2019 Red Sox sample mean for runs scored
= 5.56 (from Minitab output)

X-bar₂ = the 2019 Yankees sample mean for runs scored
= 5.82 (from Minitab output)

t* = the critical t value for our 95% confidence interval
= 1.96 (from t tables for 95%, 161 degrees of freedom)

S_1^2 = the 2019 Red Sox standard deviation squared
= 3.59^2 = 12.8881 (SD from Minitab output)

S_2^2 = the 2019 Yankees standard deviation squared
= 3.38^2 = 11.4244 (SD from Minitab output)

N₁ = the 2019 Red Sox sample size, the number of entries
= 162 (from Minitab output)

N₂ = the 2019 Yankees sample size , the number of entries
= 162 (from Minitab output)

Now I enter the numbers ...

Confidence interval: $\left(\bar{X}_1 - \bar{X}_2\right) \pm t^* \times \sqrt{\frac{s_1^2}{n_1} + \frac{s_2^2}{n_2}}$

Plug in:
$$= (5.56 - 5.82) \pm 1.96 \times \sqrt{\frac{12.8881}{162} + \frac{11.4244}{162}}$$

First we'll do our point estimate, $X\text{-bar}_1 - X\text{-bar}_2$.

Subtract:
$$= -.26 \pm 1.96 \times \sqrt{\frac{12.8881}{162} + \frac{11.4244}{162}}$$

Divide and add:
$$= -.26 \pm 1.96 \times \sqrt{.1500771605}$$

Square root:
$$= -.26 \pm 1.96 \times .3873979356$$

Multiply:
$$= -.26 \pm .7593$$

In stats the .7593 is known as our margin of error. Now to find the confidence interval, I take our point estimate, -.26 — the first number we got by subtracting $X\text{-bar}_1 - X\text{-bar}_2$ — then add and subtract the margin of error to find our high and low values.

Subtract:
$$-.26 - .7593 = -1.0193 \text{ (lower end)}$$

And **add**:
$$-.26 + .7593 = .4993 \text{ (upper end)}$$

Manually our 95% confidence interval for the 2019 Red Sox vs. Yankees = (-1.0193, .4993). We can compare better to the Minitab result, we'll round to (-1.020, .499).

Minitab calculated the confidence interval as: (-1.021, .502), manually I came pretty close to Minitab. The difference is due to me intermediate rounding while I did the math. Even though it looks like I used a lot of decimal places above, they are still rounded. We are both accurate with our intervals, but Minitab does not round till the end, making Minitab is more precise. And quicker and easier! In other words, keep on using Minitab.

In the next chapter we can go back to proportions, but expand our testing to two samples again.

2-proportions statistical test

Baseball is 90 percent mental. The other half is physical.

— *Yogi Berra, American icon*

PREVIEW: ANSWER "BUT WAS IT STATISTICALLY WORSE?" WITH ONLY TWO FRACTIONS

We are now ready to learn about 2-proportions statistical inference. A proportion (as we saw earlier) is a fraction, decimal or percent. A statistical test is when we do a hypothesis test and a 95% confidence interval (CI) to test the sample we have. And inference means making an educated guess about the population, given the sample we have.

Example: If we look up Tom Brady's passing yards in his first 4 games with Tampa in the fall of 2020 that is our sample. Then we use that sample to make an inference (guess) about his first full season with the Bucs. The full season, all 16 games, is his population in statistics.

Inference is easy. How about the proportion part? Here is an example of a 1-proportion question we did in Chapter 21.

1-proportion vs. 2-proportions differences

Springfield College's Jake Ross had a phenomenal senior season. He is in the running for Division III basketball player of the year. In his senior year, Jake made 70 out of 180 3-point shots he took. In 2018, his shooting percentage was .308 (30.8%). Did Jake's shooting statistically **improve** in 2019?

This is a 1-proportion question, because we can write Jake's shooting as 70/180 (a fraction), .3889 (a decimal which I got by dividing 70/180), or 38.9% (moving the decimal two places to the right to get a percent). It is a 1-proportion test because we only have Jake's senior year (70/180). One sample of data makes it a 1-proportion problem.

Now in Chapter 24 it looks like we are doing 2-proportion statistical inference (based on the title of the chapter). And we are. Luckily for us, we did proportions already (see Jake above) and we did 2-samples already (thank you Lamar Jackson and TB12 in Chapter 23, where we compared their seasons: two athletes, two samples).

Note: How do you tell on the exam if it is a proportion question or a mean question? On day two of the semester, we did mini-putt in class and recorded our scores. I listed all of our data on the board, and we did the mean (average) of the front nine and the back nine, and compared them. When we had a lot of numbers, the mean was a convenient method to compare.

Last chapter we listed Lamar Jackson's season in C1, and TB12's season in C2. After a full semester, the mean was still the best way to compare two seasons. We did a 2-sample means hypothesis test and confidence interval to see if there was a difference. Just like mini-putt. If you have a lot of numbers in a problem, it is a mean question. One set of numbers equals a 1-sample mean problem. Two sets of numbers equals a 2-sample means problem.

If you see no list of numbers in a problem, look for a fraction, decimal or percent (like Jake above). Then it is a proportion problem. If the problem has one proportion that equals a 1-proportion test in Minitab. If the problem has two proportions, then that equals a 2-proportion test in Minitab.

Finally, in 2-proportions testing, our test statistic is z. The same as it was in 1-proportion testing. We used t for means, but now we are back to z for proportions. In the 2-proportions answer format next, you will see a return to z for the test statistic.

2-proportions answer format

Test 1: Hypothesis Test

1. **H₀:** $P_1 = P_2$
   This is the null hypothesis, where the = sign goes. Because nothing is going on here, the 2 proportions we are investigating are assumed to be the same, they are =. We say $P_1 = P_2$ where 1 and 2 represent our 2-samples. In your answers, you will not use 1 and 2, you will talk about the problem. The 1 and 2 are here and in the alternative hypothesis just for teaching purposes.

2. **H₍A₎:** $P_1 < P_2$ or $P_1 > P_2$
   This is the alternative hypothesis, where the inequality goes. We will use less than (<) or greater than (>) depending on the wording in the problem. Are they worse? Use < Have they improved? Use >.

3. **Test statistic and its value:** z =
   We need to use z again, like we did in the 1-proportion problems. Minitab only calculates z for the 2-proportions test, making it easy for us!

4. *p*-value:
   Same probability from before that we will compare to .05.

5. **Statistical conclusion**
   Either: Since the *p*-value < .05, I reject H₀ or since the *p*-value > .05, I fail to reject H₀.

6. **Interpretation**
   If we reject H₀, there is something going on, and we have support for the H₍A₎. H₍A₎ is where we put the < or >. Now we have evidence that one team's proportion is different from the other, and in which direction. If we fail to reject, nothing is going on, it is just random variation. There is no statistical difference between the two proportions.

Test 2: Confidence Interval

7. **95% Confidence interval:**
   Just record the low and high ends of the interval here.

8. **Statistical conclusion**
   Talk about P_1 first. I am 95% confident the first proportion is less if the confidence was negative to negative, or the first proportion is more if the interval is positive to positive. Negative to positive means you are 95% confident the first proportion is between some amounts

less and some amount more. Less and more because of the negative and positive.

9. **Interpretation**

No more old number in the interval. It is the rule of signs like 2-sample means testing. If the confidence interval numbers are both negative, you would say, "There is a statistical difference, the first team is less." If the confidence interval numbers are both positive, you would say, "There is a statistical difference, the first team is more." If the confidence interval goes from negative to positive, (or the confidence interval contains 0), you would say, "There is no statistical difference between the two proportions."

Confirmation

10. **Do the results of the two tests confirm each other?**

1-6 above are the Hypothesis Test. 7-9 is the Confidence Interval test. They are both testing the same data. Do they have the same result? Look at 6 and 9 above. If they have the same result, then they confirm each other, and you answer yes. If 6 and 9 give you different answers, then they do not confirm each other, and you answer no.

GUIDED EXAMPLE FOR 2 PROPORTIONS WITH NFL KICKERS

Our 2-proportion guided example is next. Before we start it, we need the test statistic (z), *p*-value and the 95% confidence interval. We get those from Minitab. Here is the question from the 2-proportions guided example:

In the summer of 2015 the NFL moved the extra point back 13 yards. This was because year in and year out, kickers were almost automatic on extra points. But how would they do from the 15-yard line?

I looked up the last year of the old distance (2014) and I looked up 2019 to see if anything had changed. In 2014, kickers made 1222 kicks out of 1230 attempts. In 2019, they made 1136 kicks out of 1210 attempts.

Perform a hypothesis test and a confidence interval to determine if NFL kickers were statistically **better** in 2014.

Our clue for what kind of test this is, and therefore where to go on Minitab, is the "1222 kicks out of 1230 attempts". After doing means for the last two chapters, this looks more like the proportion we did in Chapter 21. Remember, a proportion (P) is a fraction, or a decimal, or a percent. The 1222 kicks out of 1230 attempts can be written as 1222/1230 (fraction), .993496 (a decimal I get by dividing 1222 by 1230 on my calculator), or 99.35% (moving the decimal over two places to get a percent).

Teacher's fun fact: Yes, that means NFL kickers made 99.35% of all extra points in 2014. An amazing number compared to today's kickers!

Our problem above is a proportion (P) problem, not a mean (M). What makes this problem different from Chapter 21 proportion problems, is this: In 2019 they made 1136 kicks out of 1210 attempts.

The 1136 out of 1210 looks like another proportion? Yes! It is another proportion. Do we have two proportions? Yes, we do. Ohhhhh, it is a 2-proportions problem similar to the 2-sample means problems we did in Chapter 23. Here though, in Chapter 24, we are going to a 2-proportion test in Minitab.

Enter data in Minitab

Choose Stat from the menu bar then Basic Statistics. Last time you clicked on "2-Sample t". Go past that and click on 2 Proportions.

Minitab defaults to Both samples are in one column but no, I looked up the data for you, so click Summarized data. Enter the data vertically, so the first box is the events, or makes (1222) and under that is the number of trials, or total attempts (1230). Then enter the 2019 data into the Sample 2 column. The 1136 goes in the top box and the 1210 in the bottom box. When in doubt, the bigger number always goes in the bottom box.

Then click Options so we can enter the > for the H_A, (I picked > because my question said: were statistically **better** in 2014, better in 2014 means 2014 was > 2019) and below that, change the default to Pool. When in doubt, always pool. Pool equates with fun!

Then say OK, OK. Remember to tell me for the test statistic on the exam that it is z = something. Because last chapter we used t, and in Chapter 25 you will learn F!

The test statistic for proportions defaulted back to z. Great! We love z. z = 7.49. The p-value = .000 = 0. Yes, 0 is OK for a p-value. It means we are pretty comfortable with the H_A, that kickers were more accurate in 2014.

Then, to get the 95% confidence interval to show up, go back to Stat, Basic statistics, 2 Proportions, and Options. Under Options, change your H_A from > to ≠. Hit OK, OK.

For every problem we do, just changing the alternative to ≠ does the trick. The 95% confidence interval = (.040424, .068882). Now you have Step 7. As percents, these are 4% to 6.9% (remember to move the decimal point in all proportion problems).

Here is a picture from Minitab of what we just did:

In 2-proportions tests, we treat confidence intervals just like we did in 2-sample means tests, we use the sign rule:

- Positive to positive, there is a difference, the first proportion is better, or higher
- Negative to negative, there is a difference, the first proportion is worse, or lower.
- Negative to positive, there is NO difference.

Our confidence interval was positive to positive, we conclude 2014 NFL kickers made between 4% and 6.9% more kicks than their 2019 counterparts. They were better at the old distance. It is not random variation.

Now we have the Minitab results (z = 7.49, *p*-value = 0, and the 95% confidence interval was (4%, 6.9%), let's try all ten steps in the 2-proportion guided example.

Answer to guided example

 In the summer of 2015 the NFL moved the extra point back 13 yards. This was because year in and year out, kickers were almost automatic on extra points. But how would they do from the 15-yard line?

I looked up the last year of the old distance (2014) and I looked up 2019 to see if anything had changed. In 2014, kickers made 1222 kicks out of 1230 attempts. In 2019, they made 1136 kicks out of 1210 attempts.

Perform a hypothesis test and a confidence interval to determine if NFL kickers were statistically **better** in 2014.

I answered the question using the **10-step answer format**:

Test 1: Hypothesis Test

1. **H_0: $P_{2014} = P_{2019}$**
   This was called a 2-proportions problem, because in the problem were two different proportions, 1222/1230 and 1136/1210. In general for my H_0 I let $P_1 = P_2$. Here it means there is no difference for the kickers. Their accuracy is the same before and after moving the extra point back.

2. **H_A: $P_{2014} > P_{2019}$**
   I used > because better in 2014 means they made a higher percentage of their kicks in 2014.

3. **Test statistic and its value:** $z = 7.49$
   Go back up to the Minitab section to see how I got the 7.49. Yes, that is a very high test statistic. One of the highest we have seen. The higher we are with z, the further out we are on the normal curve, and the more likely we are to reject H_0. This is 7.49 standard deviations above the mean. Not even Ted Williams was out here!

4. **p-value:** .000 = 0
   Yes, 0 is OK for our p-value. It means there is basically a 0% chance that kickers are the same in 2019, given our data. In your Minitab print out, there are two p-values. If they are different, use the Normal Approximation p-value.

5. **Statistical conclusion:** Since the p-value is < .05, I reject H_0.

6. **Interpretation:** NFL kickers are statistically better in 2014.
   The H_0 said they were the same. We rejected the H_0, we rejected the concept kickers were the same. The H_A is all that's left, and the H_A said they were better. We have support for the H_A here.

Test 2: Confidence Interval

7. **95% Confidence interval:** (.040424, .068882)
    To get the confidence interval in Minitab, I go back and in Options, I change the H_A from > to ≠ (not equal to), and hit OK, OK

    Also, we like means, because we can just talk about the numbers as rebounds, or pounds, or points, etc. Proportions, though, we have to change back to percents. My confidence interval above is about 4% and 7%. I use that, and the fact I put 2014 in first to talk about 2014 first, in 8:

8. **Statistical conclusion:** I am 95% confident that NFL kickers made between 4% and 7% more kicks in 2014 compared to 2019

    Both numbers were positive so I used "more" in my description.

9. **Interpretation:** Since the confidence interval is positive to positive, there is a difference, kickers were statistically better in 2014.

    Same as the last chapter, positive to positive there is a difference, the first one was higher. Negative to negative there is a difference, the first is lower. Negative to positive there is NO difference.

Confirmation

10. **Do the results of the two tests confirm each other?** Yes, the confidence interval and the hypothesis test both show there was a difference, kickers were statistically better in 2014 with the shorter distance.

Answer without teaching notes

The guided example we just did includes my notes for teaching purposes. Here is all you need for your answers on the exam:

Test 1: Hypothesis Test

1. H_0: $P_{2014} = P_{2019}$
2. H_A: $P_{2014} > P_{2019}$
3. **Test statistic and its value:** $z = 7.49$
4. ***p*-value:** .000 = 0
5. **Statistical conclusion:** Since the *p*-value is < .05, I reject H_0.
6. **Interpretation:** NFL kickers are statistically better in 2014.

Test 2: Confidence Interval

7. **95% Confidence interval:** (.040424, .068882)

8. **Statistical conclusion:** I am 95% confident that NFL kickers made between 4% and 7% more kicks in 2014 compared to 2019

9. **Interpretation:** Since the confidence interval is positive to positive, there is a difference, kickers were statistically better in 2014.

Confirmation

10. **Do the results of the two tests confirm each other?** Yes, the confidence interval and the hypothesis test both show there was a difference, kickers were statistically better in 2014 with the shorter distance.

PRACTICE PROBLEMS

In 2018 LSU was a good football team. In 2019, they were the best football team ever. And it was because of Joe Brady, the passing coordinator for the Tigers! Yes, not Joe Burrow, it was Joe Brady who made the difference between 2018 and 2019. But since Joe Burrow executed the plan perfectly, let's look at Burrow's performance.

In 2018, Joe Burrow completed 219 passes out of 379 attempts. In 2019 (with new passing game coordinator Joe Brady to help), Burrow completed 402 passes out of his 527 attempts.

Do a hypothesis test and a 95% confidence interval to determine if the 2018 version of Joe Burrow was statistically **worse** than the 2019 version.

Lamar Jackson and Tom Brady again. Last time we looked at passing yards, this time we will do completion percentage. In the 2019 season, Lamar completed 265 passes out of 401 attempts. TB12 completed 373 passes out of 613 attempts.

Do a hypothesis test and a 95% confidence interval to determine if Lamar Jackson had the statistically **better** season this time.

The Bruins came within one game of winning it all last year. This year they were poised to go all the way with the NHL's best record. Let's compare records. In 2018 the Bruins won 49 games out of the 82 they played. In 2019, the Bruins won 44 games out of 70 total games.

Do a hypothesis test and a 95% confidence interval to determine if the Bruins were statistically **worse** in 2018.

Answers to practice problems

In 2018 LSU was a good football team. In 2019, they were the best college football team ever. And it was because of Joe Brady, the passing coordinator for the Tigers! Yes, not Joe Burrow, it was Joe Brady who made the difference between 2018 and 2019. But since Joe Burrow executed the plan perfectly, let's look at Burrow's performance.

In 2018, Joe Burrow completed 219 passes out of 379 attempts. In 2019 (with new passing game coordinator Joe Brady to help), Burrow completed 402 passes out of his 527 attempts.

Do a hypothesis test and a 95% confidence interval to determine if the 2018 version of Joe Burrow was statistically **worse** than the 2019.

Test 1: Hypothesis Test

Test and CI for Two Pr...   ˅ ✕

WORKSHEET 12

Test and CI for Two Proportions

Test

Null hypothesis	H_0: $p_1 - p_2 = 0$
Alternative hypothesis	H_1: $p_1 - p_2 < 0$

Method	Z-Value	P-Value
Normal approximation	-5.91	0.000
Fisher's exact		0.000

1. **H$_0$:** $P_{2018 \text{ Joe Burrow}} = P_{2019 \text{ Joe Burrow}}$

2. **H$_A$:** $P_{2018 \text{ Joe Burrow}} < P_{2019 \text{ Joe Burrow}}$

3. **Test statistic and its value:** $z = -5.91$

4. **p-value:** .000 = 0

5. **Statistical conclusion:** Since the p-value < .05, I reject H$_0$

6. **Interpretation:** Joe Burrow was statistically worse in 2018 compared to 2019.

Test 2: Confidence Interval

Test and CI for Two Pr...   ⌄ ✕

WORKSHEET 12
Test and CI for Two Proportions

Descriptive Statistics

Sample	N	Event	Sample p
Sample 1	379	219	0.577836
Sample 2	527	402	0.762808

Estimation for Difference

Difference	95% CI for Difference
-0.184972	(-0.246546, -0.123398)

CI based on normal approximation

7. **95% Confidence interval:** (-.246546, -.123398)

8. **Statistical conclusion:** I am 95% confident that 2018 Joe Burrow completed between 25% and 12% fewer passes than the 2019 version.

9. **Interpretation:** Since the confidence interval goes from negative to negative, there is a difference, the 2018 Joe Burrow was statistically worse.

Confirmation

10. **Do the results of the two tests confirm each other?** Yes, both tests show the 2018 version of Joe Burrow was worse than his 2019 version.

Lamar Jackson and Tom Brady again. Last time we looked at passing yards, this time we will do completion percentage. In the 2019 season, Lamar completed 265 passes out of 401 attempts. TB12 completed 373 passes out of 613 attempts.

Do a hypothesis test and a 95% confidence interval to determine if Lamar Jackson had the statistically **better** season this time

Test 1: Hypothesis Test

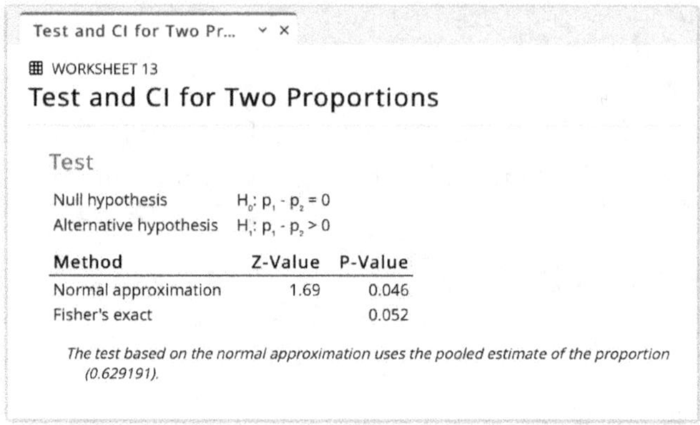

1. **H_0: $P_{\text{Lamar Jackson}} = P_{\text{Tom Brady}}$**

2. **H_A: $P_{\text{Lamar Jackson}} > P_{\text{Tom Brady}}$**

3. **Test statistic and its value:** $z = 1.69$

4. *p*-value: .046

5. **Statistical conclusion:** Since the *p*-value < .05, I reject H_0

6. **Interpretation:** Lamar Jackson had a statistically better season than Tom.

Test 2: Confidence Interval

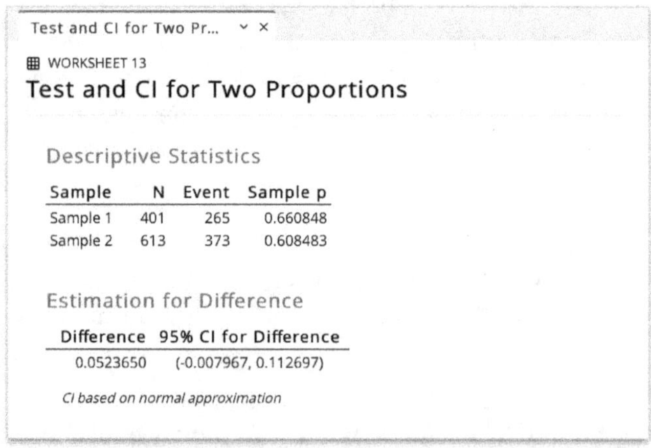

7. **95% Confidence interval:** (-.007967, .112697)

8. **Statistical conclusion:** I am 95% confident that Lamar Jackson completed between .8% fewer and 11% more passes than Tom Brady.

9. **Interpretation:** Since the interval goes from negative to positive, there is no difference in completion percentage between Lamar and TB12. Lamar was NOT having the better season. The differences we see are just random variation.

Confirmation

10. **Do the results of the two tests confirm each other?** No, the hypothesis test shows Lamar was statistically better, and the confidence interval says Lamar was NOT better.

The Bruins came within one game of winning it all last year. This year they were poised to go all the way with the NHL's best record. Let's compare records. In 2018 the Bruins won 49 games out of the 82 they played. In 2019, the Bruins won 44 games out of 70 total games.

Do a hypothesis test and a 95% confidence interval to determine if the Bruins were statistically **worse** in 2018.

Test 1: Hypothesis Test

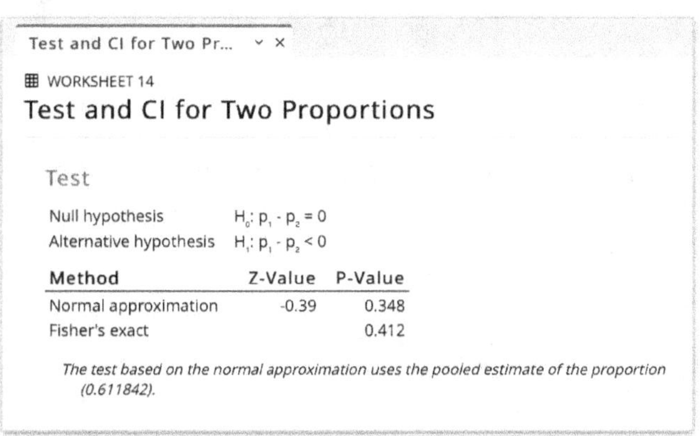

Test and CI for Two Pr...   ⌄ ✕

WORKSHEET 14

Test and CI for Two Proportions

Test

Null hypothesis	H_0: $p_1 - p_2 = 0$
Alternative hypothesis	H_1: $p_1 - p_2 < 0$

Method	Z-Value	P-Value
Normal approximation	-0.39	0.348
Fisher's exact		0.412

The test based on the normal approximation uses the pooled estimate of the proportion (0.611842).

1. **H₀:** $P_{\text{2018 Bruins}} = P_{\text{2019 Bruins}}$

2. **H_A:** $P_{\text{2018 Bruins}} < P_{\text{2019 Bruins}}$

3. **Test statistic and its value:** z = -.39

4. **p-value:** .348

5. **Statistical conclusion:** Since the *p*-value > .05, I fail to reject H₀

6. **Interpretation:** The 2018 Bruins were not statistically worse, the differences we see are just random variation.

Test 2: Confidence Interval

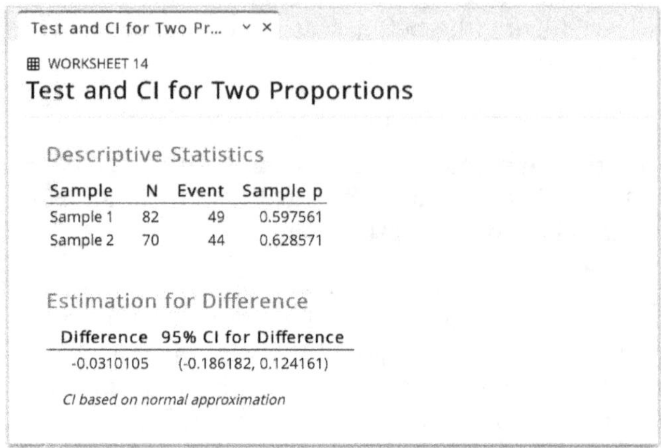

7. **95% Confidence interval:** (-.186182, .124161)

8. **Statistical conclusion:** I am 95% confident the 2018 Bruins won between 19% fewer and 12% more games than this year's team.

9. **Interpretation:** Since the confidence interval goes from negative to positive, there is no statistical difference. The 2018 Bruins were not statistically worse.

Confirmation

10. **Do the results of the two tests confirm each other?** Yes, both the confidence interval and the hypothesis test show the 2018 Bruins were not statistically worse.

BONUS PROBLEM WITH BRUCE CASSIDY AND BILL BELICHICK

Bruce Cassidy is the very successful coach of the Boston Bruins. He had the Bruins within one game of a Stanley Cup in 2019, and then the Bruins were the best team in the NHL in 2020. But he gets lost in a city dominated by the Patriots, and their infamously good head coach, Bill Belichick. Belichick may go down as the top coach of all time if he can find a way to win without Tom.

In terms of record, is Cassidy less successful than Belichick? Bruce Cassidy has won 161 games out of the 261 games he has coached. Bill Belichick has won 224 games out of 304 total. Perform a hypothesis test and a 95% confidence interval to determine if Bruce Cassidy has been statistically **less** successful than Bill Belichick.

Answer to bonus problem

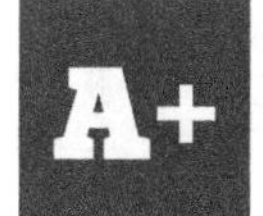

Test 1: Hypothesis Test

1. H_0: $P_{\text{Bruce Cassidy}} = P_{\text{Bill Belichick}}$

2. H_A: $P_{\text{Bruce Cassidy}} < P_{\text{Bill Belichick}}$

3. **Test statistic and its value:** $z = -3.05$

4. **_p_-value:** .0011

5. **Statistical conclusion:** Since the _p_-value < .05, I reject H_0

6. **Interpretation:** Bruce Cassidy has been statistically less successful than Bill Belichick.

Test 2: Confidence Interval

7. **95% Confidence interval:** (-.197, -.043)

8. **Statistical conclusion:** I am 95% confident Bruce Cassidy has won between 4% and 20% fewer games than Bill Belichick.

9. **Interpretation:** Since the interval goes from negative to negative, there is a difference, Bruce Cassidy has been statistically less successful than Bill Belichick.

Confirmation

10. **Do the results of the two tests confirm each other?** Yes, both the confidence interval and the hypothesis test show that Bruce Cassidy has been statistically less successful than Bill Belichick.

OPTIONAL: CONFIDENCE INTERVALS BY HAND WITH NFL KICKERS

Let's go back to our guided example:

In the summer of 2015 the NFL moved the extra point back 13 yards. This was because year in and year out, kickers were almost automatic on extra points. But how would they do from the 15 yard line?

I looked up the last year of the old distance (2014) and I looked up 2019 to see if anything had changed. In 2014, kickers made 1222 kicks out of 1230 attempts. Last year (2019) they made 1136 kicks out of 1210 attempts.

Perform a hypothesis test and a 95% confidence interval to determine if NFL kickers were statistically **better** in 2014.

We used Minitab to get the 95% confidence interval of: (.040424, .068882). See the bottom of the Minitab output:

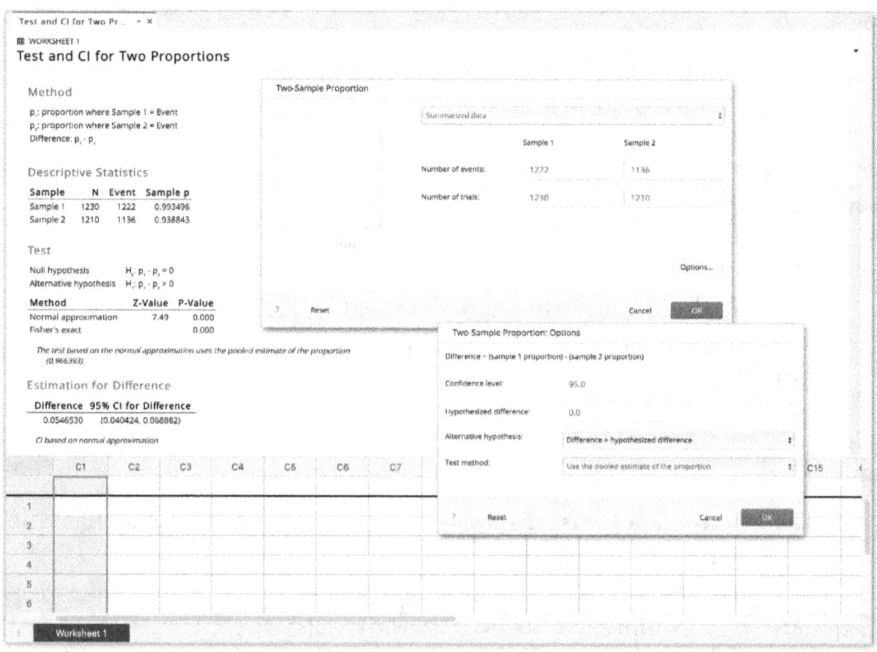

Let's see if we can get that same answer using the confidence interval formula and calculating the interval by hand. It looks like this:

$$(\hat{p}_1 - \hat{p}_2) \pm z^* \times \sqrt{\frac{\hat{p}_1 \times (1 - \hat{p}_1)}{n_1} + \frac{\hat{p}_2 \times (1 - \hat{p}_2)}{n_2}}$$

Point ±   **Margin of Error**
Estimate

The formula looks daunting! But when we break it down piece by piece it's a lot less scary.

First, all the variables labeled with a 1 represent NFL kickers in 2014. All the variables with a 2 represent NFL kickers in 2019. If it helps, the 1s and 2s can be replaced with years or any notation that makes sense to you. You can always do that if it helps you see a formula better. (Though Minitab will stubbornly insist on using 1 and 2!)

$$(\hat{p}_{2014} - \hat{p}_{2019}) \pm z^* \times \sqrt{\frac{\hat{p}_{2014} \times (1 - \hat{p}_{2014})}{n_{2014}} + \frac{\hat{p}_{2019} \times (1 - \hat{p}_{2019})}{n_{2019}}}$$

Now we can identify each part of the formula from the guided example question on NFL kickers:

n_{2014} = 1230
This is the sample size for 2014, how many extra points they attempted.

n_{2019} = 1210
This is the sample size for 2019, how many extra points they attempted.

p-hat$_{2014}$ = 1222/1230 = .993495935 (Called Sample p₁ in Minitab)
The 2014 successes over attempts. This is the success rate of our sample, the 2014 kickers. A bit over 99% success rate!

p-hat$_{2019}$ = 1136/1210 = .938842975 (Sample p₂ in Minitab)
The 2019 successes over attempts. This is the success rate of our sample, the 2019 kickers. A bit under 94% success rate.

z^* = 1.96
This is our critical z value from the z tables and represents 95% confidence.

(1 − p-hat$_{2014}$) = .006504065 just by subtracting p-hat from 1

(1 − p-hat$_{2019}$) = .061157025 just by subtracting p-hat from 1

When I did the confidence interval by hand in Chapter 23, I did intermediate rounding. In this problem, I'll keep all the decimals on my calculator throughout the problem, no intermediate rounding. I will only round at the end.

The confidence is defined as our point estimate (our "best guess") +/- the margin of error.

Now we can start plugging in values.

Do the **point estimate** part first.

Enter the values: $\left(\hat{p}_1 - \hat{p}_2\right)$ = .993495935 - .938842975 = .054652960

Do the **margin of error** part next.

$$z^* \times \sqrt{\frac{\hat{p}_{2014} \times \left(1 - \hat{p}_{2014}\right)}{n_{2014}} + \frac{\hat{p}_{2019} \times \left(1 - \hat{p}_{2019}\right)}{n_{2019}}} =$$

$$1.96 \times \sqrt{\frac{.993495935 \times .006504065}{1230} + \frac{.938842975 \times .061157025}{1210}}$$

Multiply: $\quad = 1.96 \times \sqrt{\frac{.006461762}{1230} + \frac{.057416843}{1210}}$

Divide: $\quad = 1.96 \times \sqrt{.000005253 + .000047452}$

Add: $\quad = 1.96 \times \sqrt{.000052705}$

Square root: $= 1.96 \times .007259849$

Multiply: $\quad = .014229303$

Combine the point estimate and +/- the margin of error to get our confidence interval:

Point estimate $\quad = .054652960$
Margin of error $\quad = .014229303$

Subtract: $\quad .054652960 - .014229303 = \mathbf{.040423657}$ **Low end**
and **add:** $\quad .054652960 + .014229303 = \mathbf{.068882263}$ **High end**

Minitab calculated a 95% confidence interval of: (.040424, .068882). My interval by hand rounds to (.040424, .068882). This time both intervals are exactly the same. Keeping all the decimal places helped keep my answer closer to Minitab's answer. But as always, Minitab was easier, quicker, and more precise.

We will now apply everything we have learned to ANOVA.

ANOVA statistical test

I don't believe in statistics. There are too many factors that can't be measured. You can't measure a ballplayer's heart.

— Red Auerbach, Hall of Fame basketball coach

PREVIEW: WHICH SPORT WILL PAY YOU THE MOST MONEY TO PLAY?

In the last 4 chapters we have done inference with:

1 proportion

2 proportions

1-sample mean and

2-sample means

We have tested every possible case, right? There can't be any more, can there? But there is one more case. How about 3 means? 4 means? 10 means? 15 means? 30 means? You might think that is ridiculous. No one is going to compare even 15 means right? Well, there are 15 AL teams, what if we just wanted to compare home runs for the AL teams? That would be 15 means. There are also 30 teams in MLB. What if we want to compare the average home runs hit by each team? That would be 30 means.

OK, you think, I convinced you. But the statistics must be hard. No, just as easy as the last 4 chapters. It probably takes forever, right? No, same amount of time.

Let's see an example. Springfield College is NCAA Division III, American International College (AIC) is Division II, and UMass is Division I. If we compare the heights of their basketball teams, they should be different, right? We might expect the Division I schools to be taller. And they are. But, are they statistically taller? Or is it random variation. This becomes a 3-sample mean problem. But Minitab had no option for 3-sample mean testing, all we saw was 1-sample and 2-sample. Do we have to punt? No.

To answer the question, "Is there a statistical difference in heights of the three basketball teams?" we will do our hypothesis test and 95% confidence intervals with **ANOVA**, which stands for **AN**alysis **O**f **VA**riance. Variance we saw earlier when we learned how to describe normal distributions using means and standard deviations. In our formula to calculate the standard deviation, the second to last step added a column together. That sum was called the variance! Then we took the square root of the variance to get the standard deviation.

ANOVA is what we'll use when we have three or more means to compare.

When we analyze the variance we look at the differences of the mean of each of our samples and between each sample. Then ANOVA compares those means statistically to determine if there is any difference between them.

For example, there are 15 AL teams, we could find the number of home runs each team hit in 2020. When we enter the home runs into Minitab (using C1 through C15), we notice the number of home runs hit by each team is different. When we run ANOVA, Minitab will compare all 15 teams to see if there is a statistical difference between them. But isn't there always a statistical difference? No, there is usually a difference, but it could be just random variation. If the differences we see are just random variation, then it is not a statistical difference.

F: our new test statistic

For our proportion problems, our test statistic was z. For the sample mean problems, our test statistic was t. For ANOVA, the test statistic is F. We will use F to determine if there are any differences between the means we are investigating. In case you were wondering, F is equal to the variation among the sample means divided by the variation among individuals in each sample. Lots of variation, right? Then Minitab will analyze that variation, and calculate our F.

	1 sample		2 sample		3 or more
	mean	proportion	mean	proportion	mean
Confidence interval	t	z	t	z	F
Hypothesis test	t	z	t	z	F

The larger the F value, the more likely there is a difference. The larger the F value, the smaller the p-value. How do we know if there is a statistical difference? Same as before, if the p-value $<$ .05, there is a difference. But the hypothesis test will only tell us if there is a difference. Not what the difference is.

To tell which one is bigger, or smaller, we look at the confidence intervals. If the confidence intervals overlap, there is no difference and you can draw a line through them. If at least one confidence interval does not overlap, we will conclude there is a difference, and we will know which mean is bigger or smaller. If the mean is bigger, the confidence interval is above the others. If smaller, the confidence interval is below the others.

Minitab step-by-step for ANOVA

Minitab will calculate our test statistic, p-value and 95% confidence intervals's for us. List your data in the columns in Minitab, using a different column for each group. Our example is to compare the heights of three college basketball teams (SC vs. AIC vs. UMass). We want to see if there is a statistical difference in heights. If there is a statistical difference, then we want to know which team is taller (or shorter). For our guided example, I will do the 2020 women's teams' heights. Then in your first practice problem, you will do the 2020 men's teams' heights.

Enter data in Minitab:
Label C1 2020 SC women's heights, C2 2020 AIC women's heights, and C3 2020 UMass women's heights. I went to each college's website and looked up the heights in inches for you (you are welcome!).

You will enter the 2020 heights in inches into Minitab as individual columns.

SC women's heights	AIC women's heights	UMass women's heights
73	64	69
65	64	67
65	65	71
70	69	66
66	69	67
67	72	71
67	69	74
73	68	66
66	70	71
68	68	74
66	69	73
72	65	77
60	67	73
70	73	
70		

Choose Minitab options

Pick Stat from the menu bar as usual, but go straight down to ANOVA and stop there. There are choices off to the right of ANOVA, but for us, we will keep it simple and choose One way ANOVA.

After choosing one-way ANOVA, the drop down menu defaults to data in one column. Click that box and change it because we put the data in separate columns. Then move the three columns (C1, C2, and C3) over to the response box. Then hit OK. No options or anything else to change.

Done!

What is staring at you are the three confidence intervals. Minimize them for now. Among the output in our session window, you can scroll down a little and see our test statistic, F. Here F should be 3.18. Right next to it is our p-value of .053. Then scroll down a bit and the three confidence intervals will be:

SC: (66.152, 69.582)

AIC: (66.225, 69.775)

UMass: (68.850, 72.534)

Here is a picture of the Minitab screen for ANOVA. Now we have our test statistic (F = 3.18), p-value (.053), and our three 95% confidence intervals for our guided example.

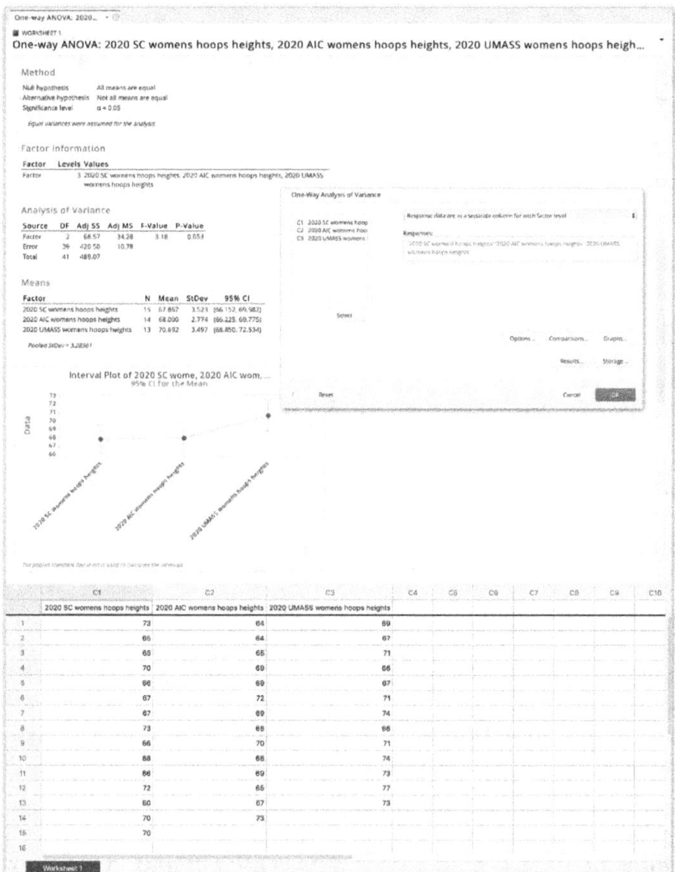

One-way ANOVA: 2020...

WORKSHEET 1

One-way ANOVA: 2020 SC womens hoops heights, 2020 AIC womens hoops heights, 2020 UMASS womens hoops heigh...

Method

Null hypothesis	All means are equal
Alternative hypothesis	Not all means are equal
Significance level	α = 0.05

Equal variances were assumed for the analysis

Factor Information

Factor	Levels	Values
Factor	3	2020 SC womens hoops heights, 2020 AIC womens hoops heights, 2020 UMASS womens hoops heights

Analysis of Variance

Source	DF	Adj SS	Adj MS	F-Value	P-Value
Factor	2	68.57	34.28	3.18	0.053
Error	39	420.50	10.78		
Total	41	489.07			

Means

Factor	N	Mean	StDev	95% CI
2020 SC womens hoops heights	15	67.867	3.523	(66.152, 69.582)
2020 AIC womens hoops heights	14	68.000	2.774	(66.225, 69.775)
2020 UMASS womens hoops heights	13	70.892	3.497	(68.850, 72.534)

Pooled StDev = 3.28361

Interval Plot of 2020 SC wome, 2020 AIC wom, ...
95% CI for the Mean

One-Way Analysis of Variance

C1 2020 SC womens hoop
C2 2020 AIC womens hoo
C3 2020 UMASS womens

Row	2020 SC womens hoops heights	2020 AIC womens hoops heights	2020 UMASS womens hoops heights
1	73	64	69
2	66	64	67
3	65	65	71
4	70	69	66
5	66	69	67
6	67	72	71
7	67	69	74
8	73	68	66
9	66	70	71
10	68	68	74
11	66	69	73
12	72	65	77
13	60	67	73
14	70	73	
15	70		
16			

Worksheet 1

GUIDED EXAMPLE FOR ANOVA WITH LSU

There are three divisions in the NCAA for college sports: Division I, II and III. Division I is the most competitive level, while Division III is more representative of the student-athlete. We would expect differences in the student-athletes at all three levels, especially height.

In our guided example, we will look at the heights of women basketball players on the 2019-20 rosters of three teams, SC, AIC and UMass. SC represents Division III, AIC Division II, and UMass Division I.

When we entered the heights into Minitab, we noticed the numbers were different. They should be different, right? Each team represents a different level of college sports.

Our question is, Are the heights statistically different or are the differences we see just random variation? To decide, we will do a hypothesis test and a 95% confidence interval and answer the question, Is there a statistical difference in heights of the three women's basketball teams?

ANOVA answer format

We will answer the question using the same 10-step answer format:

Test 1: Hypothesis Test

1. **Ho:** $M_{heights\ SC} = M_{heights\ AIC} = M_{heights\ UMass}$
   We have numbers (heights in inches) for each team, not proportions. I will use M (mean) in ANOVA. The = implies there is no difference in the mean (average) heights for the three teams.

2. **HA:** At least one mean is different
   Big change in ANOVA, there is no use of < or >. Every ANOVA problem uses "At least one mean is different" as the alternative. We treat the H_A like that because we do not know if one mean is different, or two means are different, or all means are different. If they are different, we will find out the difference in the confidence intervals.

3. **Test statistic and its value:** F = 3.18
   F is our new test statistic and was calculated previously in the Minitab ANOVA example. Also the *p*-value:

4. ***p*-value:** .053
   Right next to F in the Minitab output. It means there is a 5.3% chance there is NO difference in the heights of the three teams, given our data in the samples.

5. **Statistical conclusion:** Since the *p*-value > .05, I fail to reject H_o
   No change here, and if the *p*-value had been less than .05, we would have rejected the H_o.

6. **Interpretation:** There is no statistical difference in heights, the differences we see are just random variation.

   We tested H_o and decided we can't reject H_o. It is still there. H_o says all three means are equal, there is no statistical difference.

Test 2: Confidence Interval

7. **95% Confidence interval:**
   SC: (66.152, 69.582)
   AIC: (66.225, 69.775)
   UMass: (68.850, 72.534).

   Just list the three confidence intervals.

8. **Statistical conclusion:** I am 95% confident the SC women's basketball team heights are between 66.2 and 69.6 inches, and the AIC team is between 66.2 and 69.8 inches, and UMass women are between 68.9 and 72.5 inches tall.

   We will do the same conclusion here as we did with one mean and two means. You just have three to mention this time. And unlike proportions where you have to change the answer to a percent, here we just leave it as inches!

9. **Interpretation:** Since the confidence intervals overlap, there is no statistical difference in heights between the three teams.

   Look up at the three confidence intervals, and you can see both SC and AIC go out to 69+ inches, while UMass starts at 68.85. That is called overlapping. If confidence intervals overlap, there is no statistical difference. If they do not overlap, there is a difference.

Confirmation

10. **Do the results of the two tests confirm each other?** Yes! Both the confidence interval and the hypothesis test show there is no statistical difference in heights for the three teams.
    Confirm is when the two tests give you the same result. Even though they both show no difference, we answer yes here because the two tests gave us the same answer.

Answer without teaching notes

We are done! The answers above include my notes, here is all you need to submit.

Test 1: Hypothesis Test

1. **H₀:** $M_{\text{height SC}} = M_{\text{heights AIC}} = M_{\text{heights UMass}}$

2. **Hₐ:** At least one mean is different.

3. **Test statistic and its value:** $F = 3.18$

4. ***p*-value:** .053

5. **Statistical conclusion:** Since the p-value > .05, I fail to reject H_o

6. **Interpretation:** There is no statistical difference in heights, the differences we see are just random variation.

Test 2: Confidence Interval

7. **95% Confidence interval:**
   SC: (66.152, 69.582)
   AIC: (66.225, 69.775)
   UMass: (68.850, 72.534).

8. **Statistical conclusion:** I am 95% confident the SC women's basketball team heights are between 66.2 and 69.6 inches, and the AIC team is between 66.2 and 69.8 inches, and UMass women are between 68.9 and 72.5 inches tall.

9. **Interpretation:** Since the confidence intervals overlap, there is no statistical difference in heights between the three teams.

Confirmation

10. **Do the results of the two tests confirm each other?** Yes! Both the confidence interval and the hypothesis test show there is no statistical difference in heights for the three teams.

Confidence interval detail

Here are the confidence intervals from above:

SC: (66.152, 69.582)
AIC: (66.225, 69.775)
UMass: (68.850, 72.534)

We decided that because UMass started at 68 and SC and AIC both went over 69, that the confidence intervals overlapped. Some students would rather look at the graphs to make this decision. Earlier you minimized the confidence intervals graphs, go back and maximize it now. The trick with overlapping is if you can you draw a horizontal line through all three confidence intervals, they overlap. Here you can! As you can see in the chart below, there is a narrow strip between the lowest high bar and the highest low bar that passes through all three intervals.

But if you can't draw a line through the three confidence intervals then there is a statistical difference, and you should mention which is higher or lower.

I will show you how that works with our three intervals. For teaching purposes, I will change one number, the UMass low to 70 inches. Here are the three confidence intervals again with that one change:

SC: (66.152, 69.582)

AIC: (66.225, 69.775)

UMass: (70.000, 72.534)

Now, UMass does not start until 70 inches, and SC and AIC end at 69+, they do NOT overlap. There is a gap. In the second chart, the strip between the lowest high bar (SC) and the highest low bar (UMass) passes only through AIC. Our answer would have been, "There is a statistical difference, and that difference is the UMass is taller"!

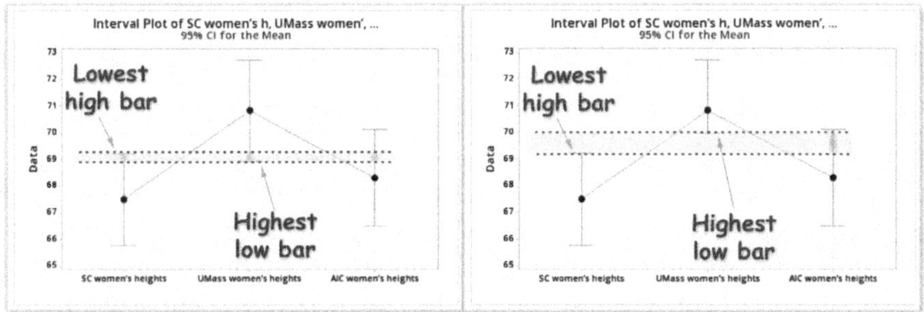

Now we can try some practice problems.

PRACTICE PROBLEMS

Two years ago, the Springfield College men's basketball team opened their season with a scrimmage against UMass, a Division I school. SC held their own despite the size difference. But is there really a statistical difference in height? I will add Division II AIC to the mix.

In Minitab, enter SC in C1, AIC in C2 and UMass in C3 (which are all in inches). Perform a hypothesis test and a 95% confidence interval to answer the question, Is there a statistical difference in heights for the men's basketball teams at the three colleges?

SC men's heights	AIC men's heights	UMass men's heights
74	79	74
72	73	74
74	73	78
75	72	75
77	69	80
70	77	77
74	76	82

SC men's heights	AIC men's heights	UMass men's heights
72	72	71
73	76	76
73	70	79
78	76	76
78	78	81
78	77	81
75	71	
	71	
	70	
	73	

You may have noticed my unhealthy obsession with LSU. Well, is it warranted? Were the 2020 National Champion LSU Tigers the best team ever? It is arguable. But, I will settle for the best LSU team over the last three years. I looked up the Tigers total yards of offense for the last three years, and listed them below.

Open Minitab and enter 2019 in C1, 2018 in C2, and 2017 in C3. Perform a hypothesis test and a confidence interval to answer the question, Is there a statistical difference in yards gained for LSU in 2017, 2018 and 2019?

2019 LSU Tigers offense	2018 LSU Tigers offense	2017 LSU Tigers offense
472	296	479
573	335	454
611	370	270
599	409	414
601	573	428
511	372	341
413	475	363
508	239	593
559	196	306
716	359	415
612	552	281
553	496	601
481	555	399
693		
631		

Q3 In 2019, the Springfield College football team was the best team in the country rushing the ball. We were ranked number one in all of Division III. Our rushing numbers were different from every other college, but were they statistically different? We will use the 2019 NEWMAC champion Worcester Polytechnic Institute (WPI) and cross-town rival Western New England University (WNEU) for comparison.

Enter SC in C1, WPI in C2, and WNEU in C3. Perform a hypothesis test and a confidence interval to answer the question, Is there a statistical difference in rushing yards in 2019 for SC, WPI and WNEU?

2019 SC rushing yards	2019 WPI rushing yards	2019 WNEU rushing yards
335	294	172
420	48	177
234	217	161
305	168	313
409	378	132
331	293	223
435	313	199
562	47	188
621	313	207
398	294	210
	229	14*

* Yes, 14!

Answers to practice problems

A1 Two years ago, the Springfield College men's basketball team opened their season with a scrimmage against UMass, a Division I school. SC held their own despite the size difference. But is their really a statistical difference in height? I will add Division II AIC to the mix.

In Minitab, enter SC in C1, AIC in C2 and UMass in C3 (which are all in inches). Perform a hypothesis test and a 95% confidence interval to answer the question, Is there a statistical difference in heights for the men's basketball teams at the three colleges?

Test 1: Hypothesis Test

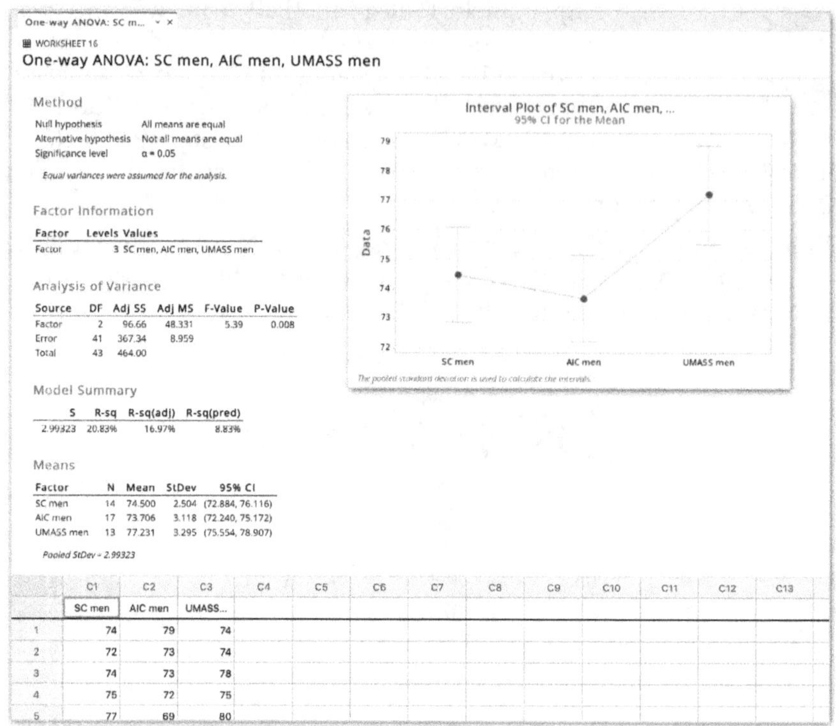

1. H_0: $M_{SC\ heights} = M_{AIC\ heights} = M_{UMass\ Heights}$

2. H_A: At least one mean is different.

3.   **Test statistic and its value:** F = 5.39

4.   **p-value:** .008

5.   **Statistical conclusion:** Since the p-value < .05, I reject H_0

6.   **Interpretation:** Yes, there is a statistical difference in heights. At least one team is different.

Test 2: Confidence Interval

7.   **95% Confidence interval:**
SC: (72.884, 76.116)
AIC: (72.240, 75.172)
UMass: (75.554, 78.907)

8.   **Statistical conclusion:** I am 95% confident that the SC men are between 72.9 and 76.1 inches tall, the AIC men are between 72.2 and 75.2 inches, and the UMass men are between 75.6 and 78.9 inches tall.

9. **Interpretation:** There is a statistical difference in heights between AIC and UMass (UMass is taller), because the confidence intervals do NOT overlap. The other intervals overlap (SC and AIC, SC and UMass) there is no difference between those teams.

Confirmation

10. **Do the results of the two tests confirm each other?** Yes, the confidence interval and the hypothesis test both show there is a statistical difference in heights, and that difference is between UMass and AIC.

You may have noticed my unhealthy obsession with LSU. Well, is it warranted? Were the 2020 National Champion LSU Tigers the best team ever? It is arguable. But, I will settle for the best LSU team over the last three years. I looked up the Tigers total yards of offense for the last three years, and listed them below.

Open Minitab and enter 2019 in C1, 2018 in C2, and 2017 in C3. Perform a hypothesis test and a confidence interval to answer the question, Is there a statistical difference in yards gained for LSU in 2017, 2018 and 2019?

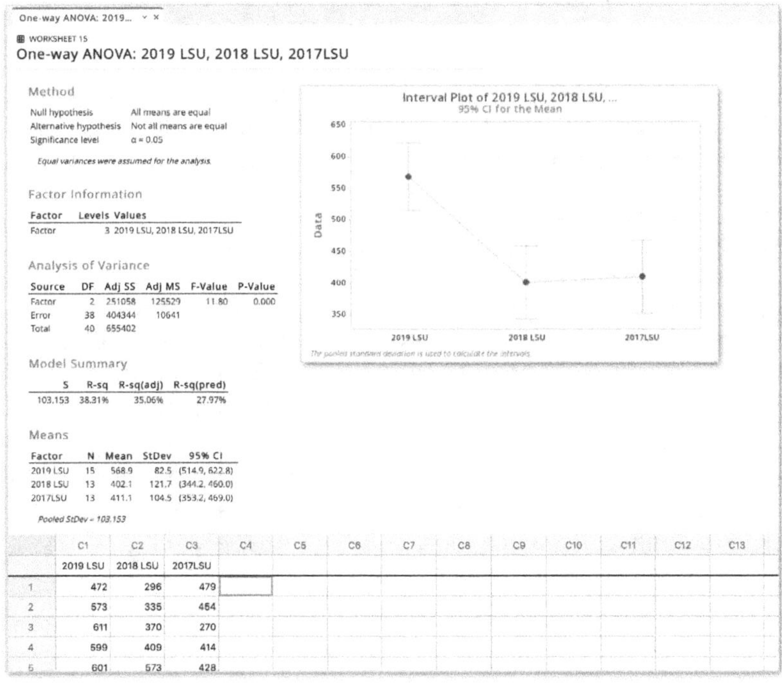

One-way ANOVA: 2019... ∨ ×

WORKSHEET 15

One-way ANOVA: 2019 LSU, 2018 LSU, 2017LSU

Method

Null hypothesis	All means are equal
Alternative hypothesis	Not all means are equal
Significance level	α = 0.05

Equal variances were assumed for the analysis.

Factor Information

Factor	Levels	Values
Factor	3	2019 LSU, 2018 LSU, 2017LSU

Analysis of Variance

Source	DF	Adj SS	Adj MS	F-Value	P-Value
Factor	2	251058	125529	11.80	0.000
Error	38	404344	10641		
Total	40	655402			

Model Summary

S	R-sq	R-sq(adj)	R-sq(pred)
103.153	38.31%	35.06%	27.97%

Means

Factor	N	Mean	StDev	95% CI
2019 LSU	15	568.9	82.5	(514.9, 622.8)
2018 LSU	13	402.1	121.7	(344.2, 460.0)
2017LSU	13	411.1	104.5	(353.2, 469.0)

Pooled StDev = 103.153

	C1	C2	C3	C4	C5	C6	C7	C8	C9	C10	C11	C12	C13
	2019 LSU	2018 LSU	2017LSU										
1	472	296	479										
2	573	335	464										
3	611	370	270										
4	599	409	414										
5	601	573	428										

Test 1: Hypothesis Test

1. **H$_0$:** M$_{2019\ LSU}$ = M$_{2018\ LSU}$ = M$_{2017\ LSU}$
2. **H$_A$:** At least one mean is different
3. **Test statistic and its value:** F = 11.80
4. ***p*-value:** .000
5. **Statistical conclusion:** Since the *p*-value < .05, I reject H$_0$
6. **Interpretation:** Yes, there is a statistical difference in yards gained. At least one team is different.

Test 2: Confidence Interval

7. **95% Confidence interval:**
   2019 LSU: (514.9, 622.8)
   2018 LSU: (344.2, 460.0)
   2017 LSU: (353.2, 469.0)
8. **Statistical conclusion:** I am 95% confident the 2019 LSU team had between 514.9 and 622.8 yards per game, the 2018 team between 344.2 and 460.0 yards per game, and the 2017 team between 353.2 and 469.0 yards per game.
9. **Interpretation:** Since the confidence intervals do not overlap, there is a statistical difference. The one team that is statistically different is the 2019 team, they had more yards per game.

Confirmation

10. **Do the results of the two tests confirm each other?** Yes, both the confidence interval and the hypothesis test show that there is a statistical difference, the 2019 LSU team was better offensively.

In 2019, the Springfield College football team was the best team in the country rushing the ball. We were ranked number one in all of Division III. Our rushing numbers were different from every other college, but were they statistically different? We will use the 2019 NEWMAC champion Worcester Polytechnic Institute (WPI) and cross-town rival Western New England University (WNEU) for comparison.

Enter SC in C1, WPI in C2, and WNEU in C3. Perform a hypothesis test and a confidence interval to answer the question, Is there a statistical difference in rushing yards in 2019 for SC, WPI and WNEU?

2019 SC rushing yards	2019 WPI rushing yards	2019 WNEU rushing yards
335	294	172
420	48	177
234	217	161
305	168	313
409	378	132
331	293	223
435	313	199
562	47	188
621	313	207
398	294	210
	229	14*

* Yes, 14!

Test 1: Hypothesis Test

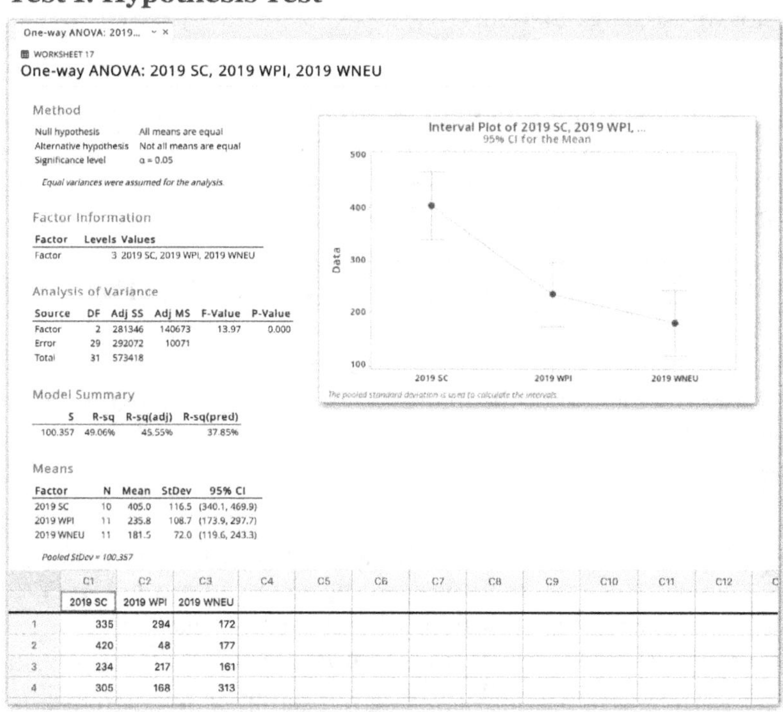

1. $\mathbf{H_0}$: $M_{SC} = M_{WPI} = M_{WNEU}$
2. $\mathbf{H_A}$: At least one mean is different.

3. **Test statistic and its value:** F = 13.97

4. ***p*-value:** .000

5. **Statistical conclusion:** Since the *p*-value < .05, I reject H_o

6. **Interpretation:** Yes, there is a statistical difference in rushing yards gained. At least one team is different.

Test 2: Confidence Interval

7. **95% Confidence interval:**
SC: (340.1, 469.9)
WPI: (173.9, 297.7)
WNEU: (119.6, 243.3)

8. **Statistical conclusion:** I am 95% confident the SC football team rushes for between 340.1 and 469.9 yards per game, WPI rushes for between 173.9 and 297.7 yards per game, and WNEU rushes for between 119.6 and 243.3 yards per game.

9. **Interpretation:** Since the confidence intervals do not overlap, there is a statistical difference. The difference is SC had more rushing yards than WPI and WNEU. WPI and WNEU overlapped, they had they were not statistically different.

Confirmation

10. **Do the results of the two tests confirm each other?** Yes, both the confidence interval and the hypothesis test show there is a statistical difference between the teams. SC rushes for yards per game than both WPI and WNEU.

BONUS PROBLEM WITH THE 4 MAJOR SPORT$

In 2020, which of the four major sports had the highest paid athletes?

There are many uses for ANOVA but we will keep it simple. We will use only the **one-way between groups ANOVA.** This will let us test the difference between groups.

For instance, we could use ANOVA to determine if there is a difference between salaries in the four major professional sports. The groups would be the player's salaries in the NFL, NBA, MLB and NHL. I can enter the salaries into Minitab, then run the one-way ANOVA to see if there are any statistical differences in the group means (the average salaries).

I copied 3277 total salaries for the four major sports and pasted them into Minitab. I pasted the NFL average salaries into C1, the NBA into C2, MLB into C3, and the NHL into C4. Then I clicked Stats, and ANOVA, then nice basic one-way.

ANOVA defaults to having the data in one column, I entered it in four different columns, so I changed that. Now in the responses box, I put all four leagues. Then I click OK. Here is the result:

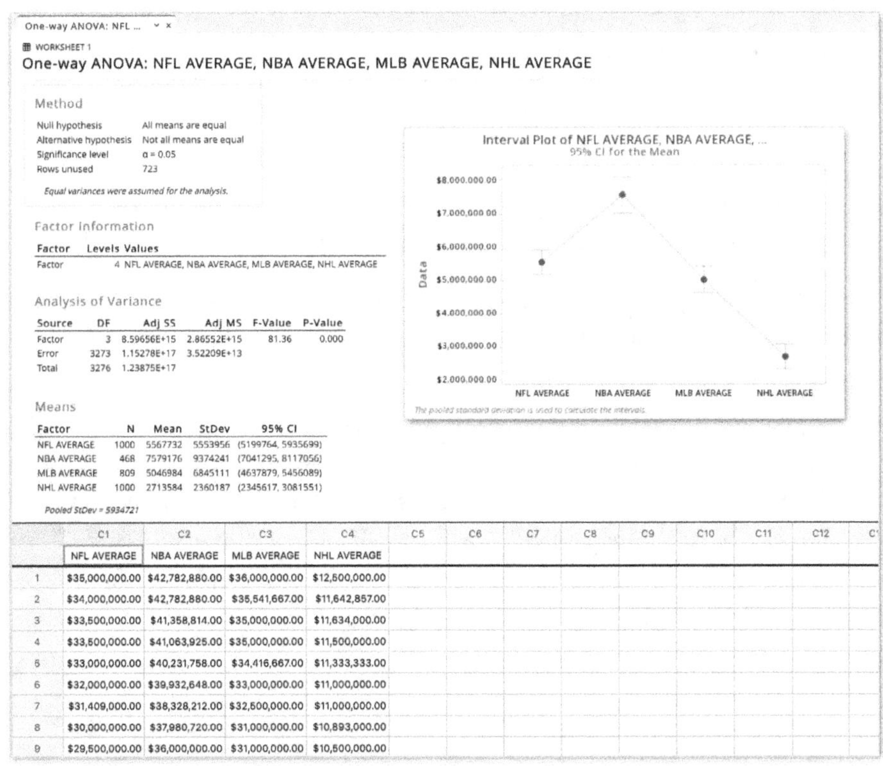

Then I performed a hypothesis test and a 95% confidence interval to determine if there was a statistical difference in salary between the four major sports (NFL, NBA, MLB, and NHL). The answer is below.

Test 1: Hypothesis Test

1. **H_0:** $M_{NFL} = M_{NBA} = M_{MLB} = M_{NHL}$

2. **H_A:** At least one of the means is different.

3. **Test statistic and its value:** $F = 81.36$

4. **p-value:** .000

5. **Statistical conclusion:** Since the p-value $< .05$, I reject H_0

6. **Interpretation:** Yes, there is a statistical difference in salaries for the four major sports. At least one sport is different.

Test 2: Confidence Interval

7. **95% Confidence interval:**
NFL: ($5,199,764, $5,935,699)
NBA: ($7,041,295, $8,117,056)
MLB: ($4,637,879, $5,456,089)
NHL: ($2,345,617, $3,081,551)

8. **Statistical conclusion:** I am 95% confident the average football player earns between $5M and $6M, the average basketball player between $7M and $8M, the average baseball player between $4.6M and $5.5M, and the average hockey player between $2.3M and $3M.

9. **Interpretation:** Since the confidence intervals do not overlap, there is a statistical difference. The difference is NBA players make the most money, and hockey players make the least money. Since the NFL and MLB overlapped, there is no statistical difference between football and baseball players.

Confirmation

10. **Do the results of the two tests confirm each other?** Yes, both the confidence interval and the hypothesis test show there is a statistical difference between leagues and salaries. NBA players make the most, NHL skaters the least. There is no statistical difference between football and baseball.

OPTIONAL: CALCULATING F BY HAND WITH THE 4 MAJOR SPORT$

Here is the formula for calculating F by hand for the bonus problem we just did.

Source of Variation	Sum of Squares	Degrees of Freedom	Mean Squares (MS)	F
Within	$SS_w = \sum_{j=1}^{k} \sum_{j=1}^{l} \left(X - \bar{X}_j \right)^2$	$df_w = k - 1$	$MS_w = \dfrac{SS_w}{df_w}$	$F = \dfrac{MS_b}{MS_w}$
Between	$SS_b = \sum_{j=1}^{k} \left(\bar{X}_j - \bar{X} \right)^2$	$df_b = n - k$	$MS_b = \dfrac{SS_b}{df_b}$	
Total	$SS_t = \sum_{j=1}^{n} \left(\bar{X}_j - \bar{X} \right)^2$	$df_t = n - 1$		

You can see above the formula for $F = \dfrac{MS_b}{MS_w}$

Let's find the top part of F first. That's the MS_b, the mean square *between* the columns of data. That's the second formula in the Mean Squares column above. To get the Sum of Squares values you *could* add all 3277 salaries by hand. Or paste them into a spreadsheet. But we'll trust Minitab's addition and plug in the summed up values and degrees of freedom from the Minitab output,

$$MS_b = \frac{SS_b}{df_b} = \frac{8.59656 \times 10^{15}}{3} = 2.86552 \times 10^{15}$$

Now, we'll find the bottom part of F, the MS_w, the mean square *within* the columns of data. That's the first formula in the Mean Squares column. Again, we'll just use the values from the Minitab output,

$$MS_w = \frac{SS_w}{df_w} = \frac{1.15278 \times 10^{17}}{3273} = 3.522089826 \times 10^{13}$$

Now substituting those into the formula for F,

$$F = \frac{MS_b}{MS_w} = \frac{2.86552 \times 10^{15}}{3.522089826 \times 10^{13}} = 81.35851558 = 81.36$$

F by hand = 81.36, which agrees with the value of F on the Minitab output of 81.36. Once again, showing us that Minitab is quicker, easier, and this time, as precise as by hand.

Teacher's note: I mentioned earlier the large amount of variance needed to calculate F and therefore do ANOVA. The within variance is calculated for each column. ANOVA calculates the variance for all the average salaries, so within each column. Then ANOVA calculates the variance between the four columns.

I entered 3277 pieces of data (salaries). ANOVA instantaneously calculates how far each salary is from the mean. Then it squares that difference for all 3277 salaries. Then it adds up all the squares, and calls it the sum of squares (SS_w). ANOVA does the same math with the four columns (the four sports), and the result is SS_b. Both those numbers are very large, so I needed to use scientific notation.

Then dividing the sum of squares values by the respective degrees of freedom (I have four columns (df = 3), and 3277 salaries (df = 3273) gives us the Mean Squares (MS_b and MS_w respectively). Then ANOVA does MS_b/MS_w to calculate F.

Practice exams for Part 3

Following each exam are answer sheets. Try the exams first before looking at the answers! Practicing all three exams will help ease your uncertainty. Practice, practice, practice and good luck!

Please answer questions using the **10-step answer format**:

Test 1: Hypothesis Test

1. H_0:
2. H_A:
3. **Test statistic and its value:**
4. *p*-value:
5. **Statistical conclusion:**
6. **Interpretation:**

Test 2: Confidence Interval

7. **95% Confidence interval:**
8. **Statistical conclusion:**
9. **Interpretation:**

Confirmation

10. **Do the results of the two tests confirm each other?**

SPRINGFIELD COLLEGE

Sports Statistics  Exam 3 Spring 2020

Q1 In 2019, the college football playoff (CFP) system worked to perfection, pitting the four best Division I college football teams against each other. They are all 13-0 (except Oklahoma which is 12-1), there is not much to separate the teams. We will look at offense through each teams' points scored, which is listed below (you do not have to look it up) and you can load into C1-C4 in Minitab. Perform a hypothesis test and a confidence interval to answer the question, Is there a difference in points scored among the four CFP teams?

OSU	LSU	Clem	OKL
45	55	52	49
42	45	24	70
51	65	41	48
76	66	52	55
48	42	21	45
34	42	45	34
52	36	45	52
38	23	59	41
73	46	59	42
56	58	55	34
28	56	52	28
56	50	38	34
34	37	62	30

Q2 Tom Brady had a bad season in 2019; his completion percentage was 29th in the NFL and basically the worst of his entire career. But do the fans in Tampa have reason to hope?

Perform a hypothesis test and a confidence interval to answer the question, Is Tom Brady statistically worse (completing a lower percentage in 2019) than 2018? In 2018, he completed 65.8% of his passes. In 2019 he completed 373 passes out of 613 attempts.

Here is the only tennis question of the year! The USTA requires the tennis balls used in the grand slam tournaments meet exact specifications. For weight, they have to be 57.6 grams. After Wilson ships the tournament balls to Flushing Meadow for the US Open, the officials do a statistical check for all requirements, including weight. For their check, they took a sample of eight balls, and weighed each. The weights are listed below, and you can load them into C5 in Minitab. The weights (in grams) were:

57.3   57.4   57.2   57.5   57.4   57.1   57.3   57.0

Those weights seem low, compared to 57.6, but are they statistically low? Or is it just random variation?

Perform a hypothesis test and a 95% confidence interval to decide if the tennis balls were too light.

The Bruins were the best team in hockey in the 2019-20 season. There were backstopped by two goalies playing at an elite level. Was there a statistical difference in the play between the two goalies? We will look at their save percentage this year. Tuukka Rask 1104 saves out of the 1181 total shots he faced. Jaroslav Halak had 832 saves out of 905 shots faced.

Perform a hypothesis test and a 95% confidence interval to determine to determine if Tuukka Rask was having the better season. (Better means he saved a higher percentage of shots.)

The 2019-20 the Celtics are one of the youngest teams in the NBA and the Lakers are the second oldest. Their ages are different. But, are they statistically different? I looked up the ages for you and listed them below. You can enter the Celtics in C6 and the Lakers' ages in C7 in Minitab.

Celtics ages		Lakers ages	
23	26	22	19
22	22	29	34
24	28	23	35
26	29	27	24
30	30	26	32
27	22	27	30
20	21	27	34
25	22	34	28
26		32	

After entering the data, perform a hypothesis test and a 95% confidence interval to determine if the Celtics are statistically younger than the Lakers.

ANSWER SHEET

Sports Statistics 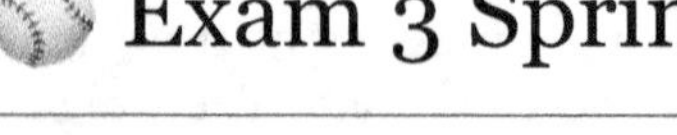 Exam 3 Spring 2020

 In 2019, the college football playoff (CFP) system worked to perfection, pitting the four best Division I college football teams against each other. They are all 13-0 (except Oklahoma which is 12-1), there isn't much to separate the teams. We will look at offense through each teams' points scored, which is listed below (you do not have to look it up) and you can load into C1-C4 in Minitab. Perform a hypothesis test and a confidence interval to answer the question, Is there a difference in points scored among the four teams?

Test 1: Hypothesis Test

1. **H$_0$:** $M_{OSU} = M_{LSU} = M_{CLEM} = M_{OKL}$
2. **H$_A$:** At least one mean is different
3. **Test statistic and its value:** F = .45
4. ***p*-value:** .720
5. **Statistical conclusion:** Since the *p*-value < .05, I fail to reject H$_0$
6. **Interpretation:** There is no statistical difference in points scored for the four teams.

Test 2: Confidence Interval

7. **95% Confidence interval:**
   OSU = (41.52, 55.86)
   LSU = (40.60, 54.94)
   CLEM = (39.37, 53.71)
   OKL = (36.06, 50.40)
8. **Statistical conclusion:** I am 95% confident that OSU will score between 41.5 and 55.9 points, LSU between 40.6 and 55 points, Clemson between 39.4 and 53.7 points and OKL between 36.1 and 50.4 points.
9. **Interpretation:** Since all four intervals overlap, there is no statistical difference in scoring between the four CFP teams.

Confirmation

10. **Do the results of the two tests confirm each other?** Yes, both tests show there is no statistical difference in scoring.

Tom Brady had a bad season in 2019; his completion percentage was 29th in the NFL and basically the worst of his entire career. Now he is getting ready to play his first games with his new team, Tampa Bay! Do the fans in Tampa have reason to hope that this year was an aberration? That Tom will be back to his normal all-world level in 2020 in a new city?

Perform a hypothesis test and a confidence interval to answer the question, Was Tom Brady statistically worse (completing a lower percentage in 2019) this year compared to last year? In 2018, he completed 65.8% of his passes. In 2019 he completed 373 passes out of 613 attempts.

Test 1: Hypothesis Test

1. **H_0:** P = .658
2. **H_A:** P < .658
3. **Test statistic and its value:** z = -2.58
4. ***p*-value:** .005
5. **Statistical conclusion:** Since the *p*-value < .05, I reject H_0
6. **Interpretation:** Tom Brady was worse this year.

Test 2: Confidence Interval

7. **95% Confidence interval:** (.569845, .647121)
8. **Statistical conclusion:** I am 95% confident Brady will complete between 57% and 64.7% of his passes.
9. **Interpretation:** Since .658 is not in the interval, it is above the interval specifically, Tom was worse this year.

Confirmation

10. **Do the results of the two tests confirm each other?** Yes, both tests show that Tom Brady was worse this year.

Here is the only tennis question of the year! The USTA requires the tennis balls used in the grand slam tournaments meet exact specifications. For weight, they have to be 57.6 grams. After Wilson shipped the tournament balls to Flushing Meadow for the US Open, the officials do a statistical check for all requirements, including weight. For their check, they took a sample of

eight balls, and weighed each. The weights are listed below, and you can load them into C5 in Minitab. The weights (in grams) were:

$$57.3 \quad 57.4 \quad 57.2 \quad 57.5 \quad 57.4 \quad 57.1 \quad 57.3 \quad 57.0$$

Those weights seem low, compared to 57.6, but are they statistically low? Or is it just random variation?

Perform a hypothesis test and a 95% confidence interval to decide if the tennis balls were too light.

Test 1: Hypothesis Test

1. **H_0:** M = 57.6
2. **H_A:** M < 57.6
3. **Test statistic and its value:** t = -5.51
4. **p-value:** .000
5. **Statistical conclusion:** Since the p-value < .05, I reject H_0
6. **Interpretation:** The tennis balls are too light.

Test 2: Confidence Interval

7. **95% Confidence interval:** (57.1355, 57.4145)
8. **Statistical conclusion:** I am 95% confident the tennis balls weigh between 57.1 and 57.4 pounds.
9. **Interpretation:** Since 57.6 is not in the interval, and the interval is specifically below the 57.6 pounds, there is a statistical difference, the tennis ball are too light.

Confirmation

10. **Do the results of the two tests confirm each other?** Yes, both tests show the same result, that the tennis balls are too light.

The Bruins were the best team in hockey in the 2019-20 season. There were backstopped by two goalies playing at an elite level. Was there a statistical difference in the play between the two goalies? We will look at their save percentage this year. Tuukka Rask 1104 saves out of the 1181 total shots he faced. Jaroslav Halak had 832 saves out of 905 shots faced.

Perform a hypothesis test and a 95% confidence interval to determine to determine if Tuukka Rask was having the better season. (Better means he saved a higher percentage of shots.)

Test 1: Hypothesis Test

1. **H_0:** $P_{Rask} = P_{Halak}$

2. **H$_A$:** P$_{Rask}$ > P$_{Halak}$

3. **Test statistic and its value:** t = 1.35

4. ***p*-value:** .088

5. **Statistical conclusion:** Since *p*-value > .05, I fail to reject H$_o$

6. **Interpretation:** There is no difference between the two goalies, Rask is not having a better season.

Test 2: Confidence Interval

7. **95% Confidence interval:** (-.007186, .038114)

8. **Statistical conclusion:** I am 95% confident that Rask is saving between .7% fewer and 3.8% more shots than Halak.

9. **Interpretation:** Since the confidence interval goes from negative to positive, there is no difference between the goalies. No difference means Rask is not having a better season.

Confirmation

10. **Do the results of the two tests confirm each other?** Yes, both tests show that Rask is not having a better season. The differences we see between the two goalies is just random variation.

The 2019-20 the Celtics are one of the youngest teams in the NBA and the Lakers are the second oldest. So their ages are different. But, are they statistically different? I looked up the ages for you and listed them below. You can enter the Celtics in C6 and the Lakers' ages in C7 in Minitab.

After entering the data, perform a hypothesis test and a 95% confidence interval to determine if the Celtics are statistically younger than the Lakers.

Test 1: Hypothesis Test

1. **H$_o$:** M$_{Celtics}$ = M$_{Lakers}$

2. **H$_A$:** M$_{Celtics}$ < M$_{Lakers}$

3. **Test statistic and its value:** t = -2.55

4. ***p*-value:** .008

5. **Statistical conclusion:** Since the *p*-value < .05, I reject H$_o$

6. **Interpretation:** The Celtics are statistically younger than the Lakers.

Test 2: Confidence Interval

7. **95% Confidence interval:** (-6.35, -.71)

8. **Statistical conclusion:** I am 95% confident that the Celtics are between 6.35 and .71 years younger than the Lakers.

9. **Interpretation:** Since the interval goes from negative to negative, there is a statistical difference. The Celtics are younger.

Confirmation

10. **Do the results of the two tests confirm each other?** Yes, both tests show that the Celtics are statistically younger than the Lakers.

SPRINGFIELD COLLEGE
Sports Statistics  Exam 3 Fall 2019

Q1 The Patriots are not clicking on offense this season, yet we already have 10 wins and should win the AFC East again. In Minitab, load the wins for all 4 teams in the AFC East from the data below (I looked up the wins for you). Over those last 12 seasons, is there a difference in wins for the 4 AFC East teams?

Pats	Dolpins	Bills	Jets
11	7	6	4
13	6	9	5
14	10	7	5
12	6	8	10
12	8	9	4
12	8	6	8
12	7	6	6
13	6	6	8
14	7	4	11
10	7	6	9
11	11	7	9
16	1	7	4

Perform a hypothesis test and a confidence interval to determine if there is a difference in wins for the Patriots, Dolphins, Bills and Jets.

Q2 Let's give some love to the Springfield Thunderbirds, the minor-league affiliate of the St. Louis Blues, who are having their best season yet. Last year they won 33 out of 76 games. So far this year they have won 14 out of 27 games. Are they statistically better than last year?

Use a hypothesis test and a confidence interval to decide if the 2019 Thunderbirds are statistically better than the 2018 team.

Q3 Joe Burrow (LSU quarterback) will win the Heisman Trophy this year. He has lead LSU's offense to historic levels, but did he step up his game? Is he playing statistically better than his 2018 season? In 2018, Burrow completed 57.8% of his passes, and so far this year he has completed 342 passes out of 439 total.

Perform a hypothesis test and a confidence interval to determine if Burrow is better in 2019.

Now let's determine if LSU's offense is playing better than last year. In C1 you can list the 2018 season for LSU for total yards gained for the team. In C2 you can list 2019. Then we can answer the question, Has LSU improved in 2019?

LSU 2018 Total Offense per game	LSU 2019 Total Offense per game
296	472
335	573
370	611
409	599
573	601
372	511
475	413
239	508
196	559
359	716
552	612
496	553
555	481

Perform a hypothesis test and a confidence interval to determine if LSU's offense is better in 2019.

Last (but not least), Springfield College has basketball-phenom Jake Ross, the best player in all Division III. He has lead the SC men's team to a 7-0 record with some of the best playing of his 4 year career. Last year he scored 23.8 PPG. Here's his scoring for 2019. You can add it to C3.

34	41	33	31	32	19	26

Using that data, perform a hypothesis test and a confidence interval to determine if Jake is playing better in 2019.

ANSWER SHEET
Sports Statistics 🏈 Exam 3 Fall 2019

 The Patriots are not clicking on offense this season, yet we already have 10 wins and should win the AFC East again. In Minitab, you loaded the wins for all 4 teams in the AFC East. Over those last 12 seasons, is there a difference in wins for the 4 AFC East teams?

Perform a hypothesis test and a confidence interval to determine if there is a statistical difference in wins for the Patriots, Dolphins, Bills and Jets.

Test 1: Hypothesis Test

1. **H_0:** $M_{Pats} = M_{Dolphins} = M_{Bills} = M_{Jets}$

2. **H_A:** At least one mean is different.

3. **Test statistic and its value:** F = 22.12

4. **p-value:** .000

5. **Statistical conclusion:** Since the p-value < .05, I reject H_0

6. **Interpretation:** There is a statistical difference in wins for the four AFC East teams.

Test 2: Confidence Interval

7. **95% Confidence interval:**
    Pats = (11.30, 13.70)
    Dolphins = (5.80, 8.20)
    Bills = (5.55, 7.96)
    Jets = (5.71, 8.12)

8. **Statistical conclusion:** I am 95% confident the Patriots will win between 11.3 and 13.7 games every year, the Dolphins between 5.8 and 8.2 games, the Bills between 5.55 and 7.96 games, and the Jets between 5.71 and 8.12 games.

9. **Interpretation:** Since the confidence intervals do not overlap, there is a statistical difference, the Patriots win more games.

Confirmation

10. **Do the results of the two tests confirm each other?** Yes, both tests show that there is a statistical difference in number of wins for the teams in the AFC Eats, and that difference is the Patriots win more games.

Let's give some love to the Springfield Thunderbirds, the minor-league affiliate of the St. Louis Blues, who are having their best season yet. Last year they won 33 out of 76 games. So far this year they have won 14 out of 27 games. Are they statistically better than last year?

Use a hypothesis test and a confidence interval to decide if the 2019 Thunderbirds are statistically better than the 2018 team.

Test 1: Hypothesis Test

1. **H_0:** $P_{2019} = P_{2018}$

2. **H_A:** $P_{2019} > P_{2018}$

3. **Test statistic and its value:** $z = .76$

4. **p-value:** .225

5. **Statistical conclusion:** Since the p-value $> .05$, I fail to reject H_0

6. **Interpretation:** The Thunderbirds are not statistically better this year.

Test 2: Confidence Interval

7. **95% Confidence interval:** (-.1346, .30326)

8. **Statistical conclusion:** I am 95% confident the 2019 Thunderbirds will win between 13.5% fewer and 30.3% more games than the 2018 team.

9. **Interpretation:** Since the interval goes from negative to positive, there is no statistical difference between the two teams. This year's team is not better.

Confirmation

10. **Do the results of the two tests confirm each other?** Yes, both tests show this year's team is not statistically better than last years!

Joe Burrow (LSU quarterback) will win the Heisman Trophy this year. He has lead LSU's offense to historic levels, but did he step up his game? Is he playing statistically better than his 2018 season? In 2018, Burrow completed 57.8% of his passes, and so far this year he has completed 342 passes out of 439 total.

Perform a hypothesis test and a confidence interval to determine if Burrow is better in 2019.

Test 1: Hypothesis Test

1. **H_0:** $P = .578$

2. **H$_A$:** P > .578

3. **Test statistic and its value:** z = 8.53

4. ***p*-value:** .000

5. **Statistical conclusion:** Since the *p*-value < .05, I reject H$_0$

6. **Interpretation:** Burrow is better in 2019.

Test 2: Confidence Interval

7. **95% Confidence interval:** (.740233, .817854) or 74% to 82%

8. **Statistical conclusion:** I am 95% confident Joe Burrow will complete between 74% and 82% of his passes.

9. **Interpretation:** Since .578 is not in the interval, and it is below the interval, Burrow is better in 2019.

Confirmation

10. **Do the results of the two tests confirm each other?** Yes, both tests show that Burrow is better in 2019.

Now let's determine if LSU's offense is playing better than last year. In C1 you can list the 2018 season for LSU for total yards gained for the team. In C2 you can list 2019. Then we can answer the question, Has LSU improved in 2019?

Perform a hypothesis test and a confidence interval to determine if LSU's offense is better in 2019.

Test 1: Hypothesis Test

1. **H$_0$:** M$_{2018}$ = M$_{2019}$

2. **H$_A$:** M$_{2018}$ < M$_{2019}$

3. **Test statistic and its value:** t = 3.80

4. ***p*-value:** .001

5. **Statistical conclusion:** Since the *p*-value < .05 I reject H$_0$

6. **Interpretation:** LSU's offense is better in 2019.

Test 2: Confidence Interval

7. **95% Confidence interval:** (-235.3, -69.6)

8. **Statistical conclusion:** I am 95% confident the 2018 LSU Tigers averaged between 235.3 and 69.6 yards per game less than the 2019 team.

9. **Interpretation:** Since the interval goes from negative to negative, there is a statistical difference, in this case the 2019 offense is better.

Confirmation

10. **Do the results of the two tests confirm each other?** Yes, both tests show the 2019 LSU offense was statistically better.

Last (but not least), Springfield College has basketball-phenom Jake Ross, the best player in all Division III. He has lead the SC men's team to a 7-0 record with some of the best playing of his 4 year career. Last year he scored 23.8 PPG. Here's his scoring for 2019. You can add it to C3.

| 34 | 41 | 33 | 31 | 32 | 19 | 26 |

Using that data, perform a hypothesis test and a confidence interval to determine if Jake is playing statistically **better** in 2019.

Test 1: Hypothesis Test

1. **H_0:** M = 23.8
2. **H_A:** M > 23.8
3. **Test statistic and its value:** t = 2.72
4. ***p*-value:** .017
5. **Statistical conclusion:** Since the *p*-value < .05, I reject H_0
6. **Interpretation:** Jake Ross is playing better in 2019.

Test 2: Confidence Interval

7. **95% Confidence interval:** (24.51, 37.21)
8. **Statistical conclusion:** I am 95% confident that Jake Ross will score between 24.51 points and 37.21 points per game in 2019.
9. **Interpretation:** Since 23.8 is not in the interval, and the interval is higher, Jake Ross is playing better.

Confirmation

10. **Do the results of the two tests confirm each other?** Yes, both tests show that Jake is playing statistically better in 2019.

SPRINGFIELD COLLEGE
Sports Statistics  Exam 3 Spring 2019

 **Q1** Springfield College gets some of the best athletes in the country to come here and play sports. We will look at the heights in inches of four SC teams to see if there is a difference in heights between the teams. Please enter the following heights in C1 – C4 in Minitab.

SC Women's Hoops	SC Men's Hoops	SC Women's VB	SC Men's VB
73	72	64	74
65	72	66	74
67	74	72	74
67	76	69	77
67	77	72	66
73	70	69	75
66	77	67	69
71	74	69	76
66	72	69	77
70	75	70	72
60	74	70	77
70	74	69	72
70	78	64	74
	80	71	78
		70	78
		68	72
		69	72
			79

Perform a hypothesis test and a confidence interval to determine if there is a statistical difference in heights for the 2019 Springfield College Women's basketball, Men's basketball, Women's volleyball and Men's volleyball teams.

 **Q2** Bill Belichick has won 225 games out of 304 with the Patriots. His biggest rival for the first decade was Peyton Manning and the Colts. Since Belichick joined the Patriots coaching staff, the Colts have won 190 out of 304. Are the Patriots better than the Colts?

Use a hypothesis test and a confidence interval to decide if the Patriots are statistically better than the Colts.

Q3 The Red Sox signed World Series hero Nathan Eovaldi to a new four-year contract last week after Eovaldi's incredible 2018 postseason. Did Eovaldi step up his game in the postseason? During the regular season his ERA was 3.81, and his postseason ERA was 1.61. It he looks like he was allowing fewer runs. But was there a statistical difference between his regular season and his postseason? His runs allowed in the regular season are: 4, 7, and 3. List them in C5, and his runs allowed in the postseason (2, 3, and 0), list them in C6. Then perform a hypothesis test and a confidence interval to determine if Eovaldi allowed fewer runs in the postseason.

Q4 Kyrie Irving had a very good shooting season this year, well above his career numbers. This season he shot 48.7%. Now in the playoffs his numbers appear to be going down, as he has made only 48 out of 114 shots.

Use a hypothesis test and a confidence interval to determine if Kyrie's shooting is statistically worse in the playoffs.

Q5 The 2018-19 Boston Bruins should to win the Stanley Cup. The scoring average to start the season was one of the best in the NHL at 3.13 goals per game. In the playoffs the Bruins have struggled and their goals scored so far are:

1	4	2	6	1	4	5	3	2	1	4

Enter the Bruins' playoff goals into C7. Then perform a hypothesis test and a confidence interval to determine if the Bruins are a worse offensive team now.

ANSWER SHEET
Sports Statistics  Exam 3 Spring 2019

 Springfield College gets some of the best athletes in the country to come here and play sports. We will look at the heights in inches of four SC teams to see if there is a difference in heights between the teams. Please enter the following heights in C1 – C4 in Minitab.

SC Women's Hoops	SC Men's Hoops	SC Women's VB	SC Men's VB
73	72	64	74
65	72	66	74
67	74	72	74
67	76	69	77
67	77	72	66
73	70	69	75
66	77	67	69
71	74	69	76
66	72	69	77
70	75	70	72
60	74	70	77
70	74	69	72
70	78	64	74
	80	71	78
		70	78
		68	72
		69	72
			79

Perform a hypothesis test and a confidence interval to determine if there is a statistical difference in heights for the 2019 Springfield College Women's basketball, Men's basketball, Women's volleyball and Men's volleyball teams.

Test 1: Hypothesis Test

1. **H_0:** $M_{\text{Womens Hoops}} = M_{\text{Mens Hoops}} = M_{\text{Womens VB}} = M_{\text{Mens VB}}$

2. **H_A:** At least one mean is different.

3. **Test statistic and its value:** F = 20.20

4. ***p*-value:** .000

5. **Statistical conclusion:** Since the p-value < .05, I reject H_0

6. **Interpretation:** There is a statistical difference in heights for the four teams.

Test 2: Confidence Interval

7. **95% Confidence interval:**
    Women's hoops = (66.394, 69.760)
    Men's hoops = (73.021, 76.264)
    Women's VB = (67.234, 70.177)
    Men's VB = (72.792, 75.652)

8. **Statistical conclusion:** I am 95% confident the women's basketball team is between 66.4 and 69.8 inches tall, the men are between 73 and 76.3 inches tall, the women's VB team is between 67.2 and 70.2 inches tall and the men are between 72.8 and 75.7 inches tall.

9. **Interpretation:** Since the confidence intervals do not overlap (there is a gap), there is a difference in heights. Specifically between the two women's teams and the two men's teams, the two men's teams are taller.

Confirmation

10. **Do the results of the two tests confirm each other?** Yes, both tests show there is a statistical difference in heights, specifically the two men's teams are taller.

Bill Belichick has won 225 games out of 304 with the Patriots. His biggest rival for the first decade was Peyton Manning and the Colts. Since Belichick joined the Patriots coaching staff, the Colts have won 190 out of 304. Are the Patriots better than the Colts?

Use a hypothesis test and a confidence interval to decide if the Patriots are statistically better than the Colts.

Test 1: Hypothesis Test

1. **H_0:** $P_{Pats} = P_{Colts}$

2. **H_A:** $P_{Pats} > P_{Colts}$

3. **Test statistic and its value:** $z = 3.05$ (remember to pool)

4. **p-value:** .001

5. **Statistical conclusion:** Since the p-value < .05, I reject H_0

6. **Interpretation:** The Patriots are statistically better than the Colts.

Test 2: Confidence Interval

7. **95% Confidence interval:** (.0417007, .188562)

8. **Statistical conclusion:** I am 95% confident the Patriots will win between 4.2% and 18.9% more games than the Colts.

9. **Interpretation:** Since the interval goes from positive to positive, there is a statistical difference, the Patriots are better.

Confirmation

10. **Do the results of the two tests confirm each other?** Yes, both tests show the Patriots are statistically better.

The Red Sox signed World Series hero Nathan Eovaldi to a new four-year contract last week after Eovaldi's incredible 2018 postseason. Did Eovaldi step up his game in the postseason? During the regular season his ERA was 3.81, and his postseason ERA was 1.61. It he looks like he was allowing fewer runs. But was there a statistical difference between his regular season and his postseason? His runs allowed in the regular season are: 4, 7, and 3. List them in C5, and his runs allowed in the postseason (2, 3, and 0), list them in C6. Then perform a hypothesis test and a confidence interval to determine if Eovaldi allowed fewer runs in the postseason.

Test 1: Hypothesis Test

1. H_0: $M_{Postseason} = M_{regular\ season}$

2. H_A: $M_{Postseason} < M_{regular\ season}$ (I used < because better in runs allowed means Eovaldi is giving up fewer runs)

3. **Test statistic and its value:** t = -2.01 (remember to assume equal variances)

4. **p-value:** .057

5. **Statistical conclusion:** Since the p-value > .05, I fail to reject H_0

6. **Interpretation:** Eovaldi did not allow fewer runs in the postseason.

Test 2: Confidence Interval

7. **95% Confidence interval:** (-7.14, 1.14)

8. **Statistical conclusion:** I am 95% confident Eovaldi gave up between 7.14 runs fewer and 1.14 runs more in the postseason.

9. **Interpretation:** Since the confidence interval goes from negative to positive, there is no difference, Eovaldi was not statistically better in the postseason.

Confirmation

10. **Do the results of the two tests confirm each other?** Yes, both tests show that Eovaldi was not statistically better in the postseason.

Kyrie Irving had a very good shooting season this year, well above his career numbers. This season he shot 48.7%. Now in the playoffs his numbers appear to be going down, as he has made only 48 out of 114 shots.

Use a hypothesis test and a confidence interval to determine if Kyrie's shooting is statistically worse in the playoffs.

Test 1: Hypothesis Test

1. **H_0:** P = .487
2. **H_A:** P < .487
3. **Test statistic and its value:** z = -1.41 (remember to use normal approximation)
4. **p-value:** .079
5. **Statistical conclusion:** Since the p-value > .05, I fail to reject H_0
6. **Interpretation:** Kyrie is not statistically worse in the playoffs.

Test 2: Confidence Interval

7. **95% Confidence interval:** (.330420, .511685)
8. **Statistical conclusion:** I am 95% confident Kyrie will make between 33% and 51% of his shots in the playoffs.
9. **Interpretation:** Since .487 is in the interval, he could still be a .487 shooter, he is not statistically worse in the playoffs.

Confirmation

10. **Do the results of the two tests confirm each other?** Yes, both tests show that Kyrie is not statistically worse in in the playoffs.

The 2018-19 Boston Bruins should to win the Stanley Cup. The scoring average to start the season was one of the best in the NHL at 3.13 goals per game. In the playoffs the Bruins have struggled and their goals scored so far are:

1 4 2 6 1 4 5 3 2 1 4

Enter the Bruins playoff goals into C7. Then perform a hypothesis test and a confidence interval to determine if the Bruins are a worse offensive team now.

Test 1: Hypothesis Test

1. **H_0:** M = 3.13
2. **H_A:** M < 3.13
3. **Test statistic and its value:** t = -.249
4. **p-value:** .404
5. **Statistical conclusion:** Since the p-value > .05, I fail to reject H_0
6. **Interpretation:** The Bruins are NOT worse now, the difference we see is just random variation.

Test 2: Confidence Interval

7. **95% Confidence interval:** (1.8364, 4.1636)
8. **Statistical conclusion:** I am 95% confident the Bruins will score between 1.8 and 4.2 goals per game in the playoffs.
9. **Interpretation:** Since 3.13 is in the confidence interval, they could still be a 3.13 goals per game team in the playoffs.

Confirmation

10. **Do the results of the two tests confirm each other?** Yes, both the confidence interval and the hypothesis test show the Bruins are not a worse offensive team in the playoffs.

And that's the ball game!

Appendix 1: Chi-Square Distribution Table

The values in the body of the table are the critical values of chi-square.

degrees of freedom	Probability of a larger value of χ^2 (p-value)									
	0.995	0.99	0.975	0.95	0.9	0.1	0.05	0.025	0.01	0.005
1	---	---	0.001	0.004	0.016	2.706	3.841	5.024	6.635	7.879
2	0.010	0.020	0.051	0.103	0.211	4.605	5.991	7.378	9.210	10.597
3	0.072	0.115	0.216	0.352	0.584	6.251	7.815	9.348	11.345	12.838
4	0.207	0.297	0.484	0.711	1.064	7.779	9.488	11.143	13.277	14.860
5	0.412	0.554	0.831	1.145	1.610	9.236	11.070	12.833	15.086	16.750
6	0.676	0.872	1.237	1.635	2.204	10.645	12.592	14.449	16.812	18.548
7	0.989	1.239	1.690	2.167	2.833	12.017	14.067	16.013	18.475	20.278
8	1.344	1.646	2.180	2.733	3.490	13.362	15.507	17.535	20.090	21.955
9	1.735	2.088	2.700	3.325	4.168	14.684	16.919	19.023	21.666	23.589
10	2.156	2.558	3.247	3.940	4.865	15.987	18.307	20.483	23.209	25.188
11	2.603	3.053	3.816	4.575	5.578	17.275	19.675	21.920	24.725	26.757
12	3.074	3.571	4.404	5.226	6.304	18.549	21.026	23.337	26.217	28.300
13	3.565	4.107	5.009	5.892	7.042	19.812	22.362	24.736	27.688	29.819
14	4.075	4.660	5.629	6.571	7.790	21.064	23.685	26.119	29.141	31.319
15	4.601	5.229	6.262	7.261	8.547	22.307	24.996	27.488	30.578	32.801
16	5.142	5.812	6.908	7.962	9.312	23.542	26.296	28.845	32.000	34.267
17	5.697	6.408	7.564	8.672	10.085	24.769	27.587	30.191	33.409	35.718
18	6.265	7.015	8.231	9.390	10.865	25.989	28.869	31.526	34.805	37.156
19	6.844	7.633	8.907	10.117	11.651	27.204	30.144	32.852	36.191	38.582
20	7.434	8.260	9.591	10.851	12.443	28.412	31.410	34.170	37.566	39.997
21	8.034	8.897	10.283	11.591	13.240	29.615	32.671	35.479	38.932	41.401
22	8.643	9.542	10.982	12.338	14.041	30.813	33.924	36.781	40.289	42.796
23	9.260	10.196	11.689	13.091	14.848	32.007	35.172	38.076	41.638	44.181
24	9.886	10.856	12.401	13.848	15.659	33.196	36.415	39.364	42.980	45.559
25	10.520	11.524	13.120	14.611	16.473	34.382	37.652	40.646	44.314	46.928
26	11.160	12.198	13.844	15.379	17.292	35.563	38.885	41.923	45.642	48.290
27	11.808	12.879	14.573	16.151	18.114	36.741	40.113	43.195	46.963	49.645
28	12.461	13.565	15.308	16.928	18.939	37.916	41.337	44.461	48.278	50.993
29	13.121	14.256	16.047	17.708	19.768	39.087	42.557	45.722	49.588	52.336
30	13.787	14.953	16.791	18.493	20.599	40.256	43.773	46.979	50.892	53.672
40	20.707	22.164	24.433	26.509	29.051	51.805	55.758	59.342	63.691	66.766
50	27.991	29.707	32.357	34.764	37.689	63.167	67.505	71.420	76.154	79.490
60	35.534	37.485	40.482	43.188	46.459	74.397	79.082	83.298	88.379	91.952
70	43.275	45.442	48.758	51.739	55.329	85.527	90.531	95.023	100.425	104.215
80	51.172	53.540	57.153	60.391	64.278	96.578	101.879	106.629	112.329	116.321
90	59.196	61.754	65.647	69.126	73.291	107.565	113.145	118.136	124.116	128.299
100	67.328	70.065	74.222	77.929	82.358	118.498	124.342	129.561	135.807	140.169